Analog
BiCMOS
DESIGN

Practices and Pitfalls

Analog BiCMOS DESIGN

Practices and Pitfalls

James C. Daly

Department of Electrical and Computer Engineering
University of Rhode Island

Denis P. Galipeau

Cherry Semiconductor Corp.

CRC Press

Boca Raton London New York Washington, D.C.

Library of Congress Cataloging-in-Publication Data

Daly, James C., 1938-
    Analog BiCMOS design : practices and pitfalls / James C. Daly, Denis P. Galipeau.
      p. cm.
    Includes bibliographical references and index.
    ISBN 0-8493-0247-1 (alk. paper)
    1. Metal oxide semiconductors, Complementary—Design and construction. 2. Bipolar
    integrated circuits—Design and construction 3. Linear integrated circuits—Design and
    construction. I. Galipeau, Denis P. II. Title.
    TK7871.9.M44 D35   1999
    621.39'732—dc21   99-047202
                                                CIP

© 2000 by CRC Press LLC

No claim to original U.S. Government works
International Standard Book Number 0-8493-0247-1
Library of Congress Card Number 99-047202
Printed in the United States of America   2 3 4 5 6 7 8 9 0
Printed on acid-free paper

Foreword

This book presents practical methods and pitfalls encountered in the design of biCMOS integrated circuits. It is intended as a reference for design engineers and as a text for an introductory course on analog integrated circuit design for engineering seniors and graduate students. A broad range of topics are covered with the intent of giving new designers the tools to complete a design project. Most of the topics have been simplified so they can be understood by students who have had a course in electronics.

The material has been used in a course open to seniors and graduate students at the University of Rhode Island. In the course, students were required to design an analog integrated circuit that was fabricated by Cherry Semiconductor Corporation.

In the process of assembling material for the book, we had discussions with many people who have been generous with information, ideas and criticism. We are grateful to James Alvernez, Mark Belch, Brad Benson, Mark Crowther, Vincenzo DiTommaso, Jeff Dumas, Paul Ferrara, Godi Fischer, Justin Fisher, Robert Fugere, Brian Harnedy, David Harrington, Ashish Kirtania, Seok-Bum Ko, Shawn LaLiberte, Andreas Ladas, Sangmok Lee, Eric Lindberg, Jien-Chung Lo, Robert Maigret, Nadia Matchey, Andrew McKinnon, Jay Moser, Ted Neira, Peter Rathfelder, Shelby Raymond, Jon Rhan, Paul Sisson, Michael Tedeschi, Claudio Tuozzolo, and Yingping Zheng.

Finally, we owe our thanks to the management and engineering staff of Cherry Semiconductor Corporation. CSC has fabricated scores of analog IC designs generated by the URI students enrolled in the course that has been the basis for this book.

James C. Daly
Denis P. Galipeau

Contents

1 **Devices**   1
   1.1 Introduction   1
   1.2 Silicon Conductivity   1
        1.2.1 Drift Current   4
        1.2.2 Energy Bands   4
        1.2.3 Sheet Resistance   7
        1.2.4 Diffusion Current   8
   1.3 Pn Junctions   9
        1.3.1 Breakdown Voltage   13
        1.3.2 Junction Capacitance   14
        1.3.3 The Law of the Junction   15
        1.3.4 Diffusion Capacitance   16
   1.4 Diode Current   17
   1.5 Bipolar Transistors   19
        1.5.1 Collector Current   20
        1.5.2 Base Current   21
        1.5.3 Ebers-Moll Model   22
        1.5.4 Breakdown   25
   1.6 MOS Transistors   26
        1.6.1 Simple MOS Model   31
   1.7 DMOS Transistors   33
   1.8 Zener Diodes   34
   1.9 EpiFETs   35
   1.10 Chapter Exercises   36

2 **Device Models**   39
   2.1 Introduction   39
   2.2 Bipolar Transistors   39
        2.2.1 Early Effect   39
        2.2.2 High Level Injection   39
        2.2.3 Gummel-Poon Model   40
   2.3 MOS Transistors   44
        2.3.1 Bipolar SPICE Implementation   49

2.4 Simple Small Signal Models for Hand
    Calculations 51
    2.4.1 Bipolar Small Signal Model 51
    2.4.2 Output Impedance 52
    2.4.3 Simple MOS Small Signal Model 53
2.5 Chapter Exercises 54

3 Current Sources   57
3.1 Current Mirrors in Bipolar Technology 59
3.2 Current Mirrors in MOS Technology 71
3.3 Chapter Exercises 76

4 Voltage References   79
4.1 Simple Voltage References 79
4.2 Vbe Multiplier 80
4.3 Zener Voltage Reference 82
4.4 Temperature Characteristics of I_c and V_{be} 83
4.5 Bandgap Voltage Reference 85

5 Amplifiers   89
5.1 The Common-Emitter Amplifier 90
5.2 The Common-Base Amplifier 96
5.3 Common-Collector Amplifiers (Emitter
    Followers) 98
5.4 Two-Transistor Amplifiers 99
5.5 CC-CE and CC-CC Amplifiers 99
5.6 The Darlington Configuration 101
5.7 The CE-CB Amplifier, or Cascode 103
5.8 Emitter-Coupled Pairs 104
5.9 The MOS Case: The Common-Source
    Amplifier 113
5.10 The CMOS Inverter 115
5.11 The Common-Source Amplifier with Source Degeneration 115
5.12 The MOS Cascode Amplifier 117
5.13 The Common-Drain (Source Follower)
    Amplifier 117
5.14 Source-Coupled Pairs 118
5.15 Chapter Exercises 123

6 Comparators   127
6.1 Hysteresis 133
    6.1.1 Hysteresis with a Resistor Divider 133
    6.1.2 Hysteresis from Transistor Current Density . . . 134
    6.1.3 Comparator with V_{be}-Dependent Hysteresis . . . 136

	6.2	The Bandgap Reference Comparator	137
	6.3	Operational Amplifiers	138
	6.4	A Programmable Current Reference	139
	6.5	A Triangle-Wave Oscillator	140
	6.6	A Four-Bit Current Summing DAC	142
	6.7	The MOS Case	143
	6.8	Chapter Exercises	143

7 Amplifier Output Stages   **145**

	7.1	The Emitter Follower: a Class A Output Stage	146
	7.2	The Common-Emitter Class A Output Stage	149
	7.3	The Class B (Push-Pull) Output	150
	7.4	The Class AB Output Stage	154
	7.5	CMOS Output Stages	154
	7.6	Overcurrent Protection	155
	7.7	Chapter Exercises	157

8 Pitfalls   **159**

	8.1	IR Drops	159
		8.1.1 The Effect of IR Drops on Current Mirrors . . .	162
	8.2	Lateral pnp	166
		8.2.1 The Saturation of Lateral pnp Transistors	166
		8.2.2 Low Beta in Large Area Lateral pnps	167
	8.3	npn Transistors	169
		8.3.1 Saturating npn Steals Base Current	169
		8.3.2 Temperature Turns On Transistors	171
	8.4	Comparators	173
		8.4.1 Headroom Failure	173
		8.4.2 Comparator Fails When Its Low Input Limit Is Exceeded	175
		8.4.3 Premature Switching	177
	8.5	Latchup	179
		8.5.1 Resistor ISO EPI Latchup	182
	8.6	Floating Tubs	185
	8.7	Parasitic MOS Transistors	187
		8.7.1 Examples of Parasitic MOSFETs	188
		8.7.2 OSFETs	188
		8.7.3 Examples of Parasitic OSFETs	189
	8.8	Metal Over Implant Resistors	190

9 Design Practices   **193**

	9.1	Matching	193
		9.1.1 Component Size	193
		9.1.2 Orientation	194

9.1.3 Temperature 195
9.1.4 Stress 195
9.1.5 Contact Placement for Matching 197
9.1.6 Buried Layer Shift 197
9.1.7 Resistor Placement 198
9.1.8 Ion Implant Resistor Conductivity Modulation . 199
9.1.9 Tub Bias Affects Resistor Match 200
9.1.10 Contact Resistance Upsets Matching 201
9.1.11 The Cross Coupled Quad Improves Matching . . 201
9.1.12 Matching Calculations 202
9.2 Electrostatic Discharge Protection (ESD) 208
9.3 ESD Protection Circuit Analysis 210
9.4 Chapter Exercises 214

Index   217

chapter 1

Devices

1.1 Introduction

The properties and performance of analog biCMOS integrated circuits are dependent on the devices used to construct them. This chapter is a review of the operation of silicon devices. It begins with a discussion of conductivity and resistance. Simple physical models for bipolar transistors, MOS transistors, and junction and diffusion capacitance are developed.

1.2 Silicon Conductivity

The conductivity of silicon can be controlled and made to vary over several orders of magnitude by adding small amounts of impurities. Silicon belongs to group four in the periodic table of elements. It has four valence electrons in its outer shell. A silicon atom in a silicon crystal has four nearest neighbors. Silicon forms covalent bonds where each atom shares its valence electrons with its four nearest neighbors. Each atom has its four original valence electrons plus the four belonging to its neighbors. That gives it eight valence electrons. The eight valence electrons complete the shell producing a stable state for the silicon atom. Electrical conductivity requires current consisting of moving electrons. The valence electrons are attached to an atom and are not free to move far from it. Some valence electrons will receive enough thermal energy to free themselves from the silicon atom. These electrons move to energy levels in a band of energy called the conduction band. Conduction band electrons are not attached to a particular atom and are free to move about the crystal.

When an electron leaves a silicon atom, the atom becomes a positively charged silicon ion. The situation is represented schematically in Figure 1.1. The vacant valence state, previously occupied by the electron, is called a hole. Each hole has a positive charge equal to one electronic charge associated with it. With one electron gone, there are seven valence electrons, shared with nearby neighbor atoms, and one hole associated with the ionized silicon atom. Holes can move. If the hole represents a missing electron that was shared with the silicon neighbor on the left, only a small amount of energy is required for one of the other seven valence electrons to move into the hole. If an electron shared with an atom on the right moves into the hole on the left, the hole will have moved from the left of the atom to the right. The movement of holes in silicon is really the movement of electrons leaving and filling electron states. It is like the motion of a bubble in a fluid. The bubble is the absence of the fluid. The bubble appears to move up, but actually the fluid is moving down. Each hole in silicon is a mobile positive charge equal to one electronic charge.

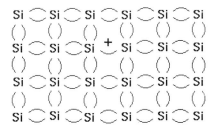

Figure 1.1   A schematic representation of a silicon crystal is shown. Each silicon atom shares its four valence electrons with its nearest neighbors. A positively charged "hole" exists where an electron has been lost due to ionization. The hole acts as a mobile positive particle with a charge equal to one electronic charge.

The conductivity of silicon increases when there are more charge carriers (electrons and holes) present. In pure silicon there will be a small number of thermally generated electron hole pairs. The number of electrons equals the number of holes because each electron leaving a silicon atom for the conduction band leaves behind a hole in the valence band. When the number of holes equals the number of conduction electrons, this is called intrinsic silicon. The intrinsic carrier concentration is strongly temperature dependent. At room temperature, the intrinsic carrier concentration $n_i = 1.5x10^{10}$ electron-hole pairs/cm^3.

Small amounts of impurity elements from group 3 or group 5 in the periodic table are used to control the electron and hole concentrations. A group five element such as phosphorus, when added to the silicon crystal replaces a silicon atom. Phosphorus has five valence electrons in its outer shell, one more than silicon. Four of phosphorus' valence electrons form covalent bonds with its four silicon neighbors. The remaining phosphorus electron is loosely associated with the phosphorus atom. Only a small amount of energy is required to ionize the phosphorus atom by moving the extra electron to the conduction band leaving behind a positively ionized phosphorus atom. Since an electron is added to the conduction band, the added group five impurity is called a donor. This represents n-type silicon with mobile electrons and fixed positively ionized donor atoms. N-type silicon is typically doped with 10^{15} or more donors per cubic centimeter. This swamps out the thermally generated electrons at normal operating temperatures.

A group three element, like boron, is called an acceptor. Doping with acceptors results in p-type silicon. When an acceptor element with three valence electrons in its outer shell replaces a silicon atom, it becomes a negative ion, acquiring an electron from the silicon. That allows it to complete its outer shell and to form covalent bonds with neighboring silicon atoms. The electron acquired from the silicon leaves a hole behind. At room temperature all acceptors are ionized and the number of holes per cubic centimeter is equal to the number of acceptor atoms.

In an n-type semiconductor, electrons are the majority carriers and holes are referred to as minority carriers. Similarly, in p-type semiconductors, holes are the majority carriers and electrons are referred to as minority carriers. In practical devices, doping levels greatly exceed the thermally generated levels of electron hole pairs (by 5 orders of magnitude or more). When silicon is doped, say, with donors to produce n-type silicon, the number of holes is reduced. The large number of electrons increases the probability of a hole recombining with an electron. An equilibrium develops where the increase of holes due to thermal generation equals the decrease of holes due to recombination. The recombination rate and the number of holes varies inversely with the number of electrons. This is called the law of mass action. It holds for all doping levels in both p-type and n-type semiconductors in equilibrium. It is a very useful relationship that allows the number of minority carriers to be calculated when the doping level for the majority carriers is known. The law of mass action is

$$pn = n_i^2 \tag{1.1}$$

where p is the number of holes per cubic cm and n is the number of conduction electrons per cubic cm.

Example

A silicon sample is doped with $N_D = 5x10^{17}$ donors/cm^3. What are the majority and minority carrier concentrations?

The sample is n-type where electrons are the majority carriers. Assuming all donors are ionized, the electron density is equal to the donor concentration, $n = 5x10^{17} cm^{-3}$. The minority (hole) concentration is

$$p = \frac{n_i^2}{N_D} = \frac{(1.5x10^{10})^2}{5x10^{17}} = 450cm^{-3}$$

There are very few holes compared to electrons in this n-type sample.

1.2.1 Drift Current

Voltage across a silicon sample results in an electric field that exerts a force on free electrons and holes causing them to move resulting in current flow. Consider an electron. The force produced by the electric field causes it to accelerate. Its velocity increases with time until it strikes the silicon crystal lattice or an impurity, where it is scattered and loses its momentum. The electron is constantly accelerating then bumping into the silicon losing its momentum. This process results in an average velocity proportional to the electric field called the drift velocity.

$$v_{drift} = \mu_n \mathcal{E} \tag{1.2}$$

where μ_n and $\mathcal{E}$ are the electron mobility and the electric field. Mobility decreases when there is more scattering of carriers. Lattice scattering increases with temperature. Therefore, mobility and conductivity tend to decrease with temperature. Carriers are also scattered from impurities. Mobility decreases significantly with doping as shown in Figure 1.2.[2]. Conductivity is proportional to mobility and carrier concentration. For an n-type sample, the current flowing through the cross-sectional area A is

$$I = Aq\mu_n \mathcal{E} = A\sigma\mathcal{E} \tag{1.3}$$

where q is the electronic charge, n is the number of free electrons per cubic centimeter, and $\sigma = q\mu_n n$ is the conductivity. Since the sample is doped with N_D donors per cubic centimeter, $n = N_D$ and the conductivity is

$$\sigma = q\mu_n N_D \tag{1.4}$$

similarly the conductivity of p-type silicon, doped with acceptor atoms, where the current carriers are holes is $\sigma = q\mu_p N_A$, where N_A is the number of acceptor atoms per cubic centimeter.

1.2.2 Energy Bands

The energy states that can be occupied by electrons are limited to bands of energy in silicon as shown in Figure 1.3. The valence band is normally

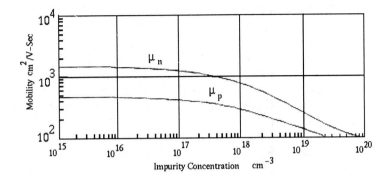

Figure 1.2   Carrier mobility in silicon at $300\,^{\circ}K$ decreases significantly with impurity concentration.[1] (Reprinted from Solid-State Electronics, Volume II, S. M. Sze and J. C. Irvin, *Resistivity, Mobility and Impurity Levels in GaAs, Ge, and Si at 300° K.*, pages 599-602, Copyright 1968, with permission from Elsevier Science.)

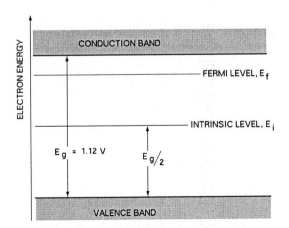

Figure 1.3   Electron energies in silicon are shown. Electrons free to move about the crystal occupy states in the conduction band. Valence electrons attached to silicon atoms occupy the valence band. The intrinsic level is approximately half way between the conduction and valence bands. The Fermi level shown corresponds to n-type silicon.

occupied by valence electrons attached to silicon atoms. The conduction band is occupied by conduction electrons that are free to move about the crystal. If all electrons are in their lowest energy states, they are occupying states in the valence band. The difference between the conduction band edge and the valence band edge is $E_G = 1.12\ eV$, the band gap. When a silicon atom loses an electron, it takes 1.12 electron volts of energy for the electron to move from the valence to the conduction band. When this happens the conduction band is occupied by an electron and the valence band is occupied by a hole. Impurities introduce electron states inside the band gap close to the valence or conduction band. Donor states are close to the conduction band. It takes very little energy for an electron to move from a donor state to the conduction band. Acceptor states are located close to the valence band. A valence electron can easily move from the valence band to an acceptor state.

The Fermi level is a measure of the probability that a state is occupied by an electron. States below the Fermi level tend to be occupied, while states above it tend to be unoccupied. As the temperature increases, some states below the Fermi level will become unoccupied as electrons move up to levels above the Fermi level. States at the Fermi level have a 50-50 chance of being occupied. In intrinsic silicon where the number of holes equals the number of electrons, the Fermi level is approximately half way between the valence and conduction bands. This Fermi level is called the intrinsic level, E_i. In an n-type semiconductor, with conduction band states occupied, the Fermi level moves up closer to the conduction band as the probability that a conduction band state is occupied increases. In p-type semiconductors, with vacant valance band states (holes), the Fermi level moves down closer to the valence band.

The position of the Fermi level relative to the intrinsic level is a measure of the carrier concentration. For n-type silicon, the Fermi level, E_f is above E_i. For p-type it is below E_i. The number of electrons per cubic cm in the conduction band is related to the position of the Fermi level by the following equation[3, page 22].

$$n = n_i e^{\frac{E_f - E_i}{KT}} \qquad (1.5)$$

where n_i is the intrinsic carrier concentration and $K = 8.62x10^{-5}$ electron volts per degrees Kelvin is Boltzmann's constant. If $T = 300$, $KT = 0.0259\ V$. 300 degrees Kelvin is 27 degrees C and 80.6°F, commonly called room temperature.

Since by the law of mass action $pn = n_i^2$

$$p = n_i e^{\frac{E_i - E_f}{KT}} \qquad (1.6)$$

Example

If a silicon sample is doped with 10^{17} acceptors per cm^3, calculate the position of the Fermi level relative to the intrinsic level at room temperature.

At normal operating temperatures, all acceptors will be ionized and the hole concentration p will equal the acceptor concentration.

$$p = N_A = 10^{17} \text{ holes per } cm^3$$

From Equation 1.6:

$$E_i - F_f = KT \, ln \left(\frac{N_A}{n_i} \right) = 0.0259 \, ln \left(\frac{10^{17}}{1.5x10^{10}} \right) = 0.41 \, V$$

The Fermi level is 0.41 V below the intrinsic level.

1.2.3   Sheet Resistance

Sheet resistance is an easily measured quantity used to characterize the doping of silicon. Consider the sample shown in Figure 1.4. The silicon is doped with donors to form a resistor of n-type silicon. The resistor length is L and its cross-sectional area is tW, where t is the effective depth of the resistor. The resistance is

$$R = \frac{L}{\sigma tW} = \frac{L}{W} R_{sh} \tag{1.7}$$

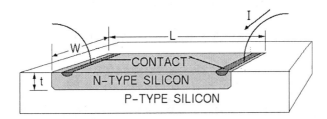

Figure 1.4   Resistors are formed in silicon by placing dopants in a specific region.

The parameter R_{sh} is the sheet resistance. Its units are ohms per square. The dimensionless quantity, L/W is the number of squares of resistive material in series between the contacts. Resistors of various values can be obtained by varying the width and length. The sheet resistance is a process parameter dependent on doping:

$$R_{sh} = \frac{1}{\sigma t} = \frac{1}{q\mu N_D t} \tag{1.8}$$

where N_D is an average doping. Usually doping varies with distance down from the surface of the silicon. $N_D t$ is the number of donors per unit area.

1.2.4 Diffusion Current

The current flow mechanism responsible for the characteristics of diodes and transistors is diffusion. Diffusion current flows without being caused by an electric field. Electrons and holes in semiconductors are in constant thermal motion. When there is a nonuniform distribution of carriers (electrons or holes), random motion causes a net motion away from the region where the electrons or holes are more dense. Consider the nonuniform distribution of holes shown in Figure 1.5. The charged particles, represented by plus signs, are equally likely to move either to the right or to the left. Because there are more particles on the left there is a net motion of one particle to the right passing across each vertical plane. This situation can exist at a pn junction, where an unlimited supply of free carriers, caused by a forward bias voltage, allows a concentration gradient to be maintained. In Figure 1.5, carrier motion is indicated by the arrows. Random motion is modeled by grouping carriers together in pairs with opposite velocities so the average velocity is zero. The overall result is the movement of one carrier from each region of high concentration to the neighboring low concentration region. If the distribution of carriers is maintained, there will be a constant current flow from left to right.

The diffusion current density for holes is given by

$$J_p = -qD_p \frac{dp}{dx} \tag{1.9}$$

where J_p is the current density, amperes/cm^2, D_p is the diffusion constant and p is the hole density, holes/cm^2.

Einstein's relation shows the diffusion constant for holes to be proportional to mobility [3, page 38]:

$$D_p = \mu_p V_T \tag{1.10}$$

and for electrons

$$D_n = \mu_n V_T \tag{1.11}$$

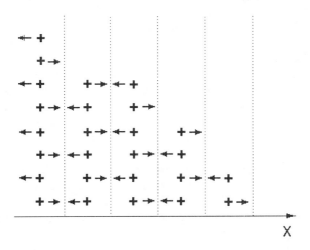

Figure 1.5 The nonuniform distribution of randomly moving positive charges results in a systematic motion of charge. Here a positive current is moving to the right.

where $V_T = KT/q$ is the thermal voltage. $V_T = 26\ mV$ at room temperature. K is Boltzmann's constant. $q = 1.6x10^{-19}\ C$ is the electronic charge and T is the absolute temperature. It is not surprising that the mobility is proportional to the diffusion constant since both describe the motion of charge in the silicon crystal.

1.3   *Pn Junctions*

Pn junctions are the building blocks of integrated circuit components. They serve as parts of active components, such as the base-emitter or collector-base junctions of a bipolar transistor, or as isolation between components, as is the case when an integrated resistor is fabricated in a reverse-biased tub. Each pn junction has a parasitic capacitance associated with it that affects device performance. Important properties such as breakdown voltage and output resistance are dependent on properties of pn junctions. Since this text isn't intended to teach device physics, we will review pn junctions only so far as is required to understand transistor operation.

Consider a pn junction under reverse bias conditions as shown in Figure 1.6, and assume that the doping is uniform in each section, with $N_D cm^{-3}$ donor atoms in the n-region and $N_A cm^{-3}$ acceptor atoms in the p-region. At the junction, there is a region devoid of electrons and holes. The electrons have moved from the n-region into the p-region where they recombine with holes. Similarly, holes move from the p-region to the n-region where they recombine with electrons. This process leaves positive donor ions in the n-region and negative acceptor ions in the p-region. The donors and acceptors occupy fixed positions in the silicon crystal and cannot move. An electric field exists between the positive donor ions in the n-region and the negative acceptor ions in the p-region. As electrons leave the n-region for the p-region, the n-region becomes positively charged and the p-region becomes negatively charged. The electric field increases until it inhibits any further movement of holes and electrons. The region near the junction devoid of charge is called the space-charge region or depletion region. An approximation that results in an accurate model of the junction is to assume the depletion region to be well defined with a definite width with an abrupt change in the carrier concentration at the edge of the depletion region. The area outside the depletion region is the charge neutral region. In the n charge-neutral region the number of negatively charged electrons equals the number of positively charged donor atoms. In the charge-neutral region in the p material the number of positively charged holes equals the number of negatively charged acceptor atoms.

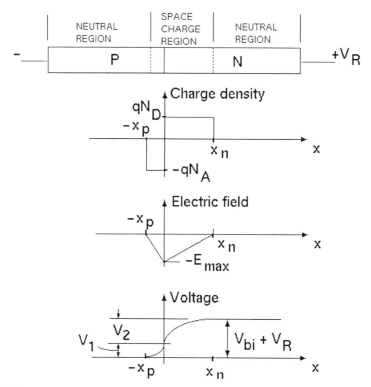

Figure 1.6   Junction charge distribution and fields.

When there is no applied bias voltage, a built-in potential, denoted Ψ, exists due to the charge distribution across the junction. This potential is just large enough to counter the diffusion of mobile charge across the junction and results in the junction being at equilibrium with no net current flow. The value of this potential is

$$\Psi = V_T \ln \left[\frac{N_A N_D}{n_i^2} \right] \tag{1.12}$$

where $V_T = kT/q$ is 26 mV at room temperature, and $n_i = 1.5x10^{15} cm^{-3}$ is the intrinsic carrier concentration of silicon.

In Figure 1.6, an applied reverse bias is added to the built-in potential, and the total voltage found across the junction is $\Psi + V_R$. If we assume the depletion region extends a distance x_p into the p-region, and distance x_n into the n-region, then

$$x_p N_A = x_n N_D \tag{1.13}$$

This is true because the charge on one side of the depletion region must be equal in magnitude and opposite in sign to the charge on the opposite side of the depletion region.

From Gauss' Law we have

$$\nabla \cdot D = \rho \tag{1.14}$$

In one dimension, this reduces to

$$\frac{dD}{dx} = \rho \tag{1.15}$$

Since $D = \epsilon \mathcal{E}$, we have

$$\frac{d\mathcal{E}}{dx} = \frac{\rho}{\epsilon} \tag{1.16}$$

Electric field can then be defined

$$\mathcal{E} = -\frac{dV}{dx} \tag{1.17}$$

Within the confines of the depletion region, the charge distribution ρ is equal to $qN_A coul/cm^3$ in the p-region, and is equal to $qN_D coul/cm^3$ in the n-region. The maximum value of the electric field across the depletion region is found at x = 0 and has a value

$$\mathcal{E}_{max} = -\frac{qN_A x_p}{\epsilon} = -\frac{qN_D x_n}{\epsilon} \tag{1.18}$$

where ϵ is the permittivity of silicon.

We have assumed the depletion region and junction boundaries are sharp and well defined. Defining the potential between $x = -x_p$ and $x = 0$ as V_1.

$$V_1 = -\int_{-x_p}^{0} \mathcal{E}\,dx = \frac{qN_A x_p^2}{2\epsilon} \tag{1.19}$$

Similarly, if we define the potential between $x = 0$ and $x = x_n$ as V_2, we obtain

$$V_2 = -\int_{0}^{x_n} \mathcal{E}\,dx = \frac{qN_D x_n^2}{2\epsilon} \tag{1.20}$$

The voltage across the depletion region is then the sum of V_1 and V_2 and may be written

$$\Psi_o + V_R = V_1 + V_2 = \frac{q}{2\epsilon}\left[N_A x_p^2 + N_D x_n^2\right] \tag{1.21}$$

Factoring and using Equation 1.13:

$$\Psi_o + V_R = \frac{qx_p^2 N_A^2}{2\epsilon}\left[\frac{1}{N_A} + \frac{1}{N_D}\right] \tag{1.22}$$

Recall Equation 1.13, $N_A x_p = N_D x_n$. If one region is much more heavily doped than the other, the depletion region exists almost entirely in the lightly doped region. This leads to an approximation called the single-sided junction. For example, if $N_A \gg N_D$, then $x_n \gg x_p$. Since the total depletion width is $x_d = x_p + x_n$, we can approximate $x_d \approx x_n$. Since $N_D \ll N_A$, Equation 1.22 becomes

$$\Psi_o + V_R = \frac{qx_n^2 N_D^2}{2\epsilon}\left[\frac{1}{N_D}\right] \tag{1.23}$$

The voltage across the junction exists across x_n, and is approximately $\Psi_o + V_R \approx V_2$.

Also, from Equations 1.23 and 1.18, the width of the depletion region and the maximum electric field are

$$x_d \approx x_n = \sqrt{\frac{2\epsilon(\Psi_o + V_R)}{qN_D}} \tag{1.24}$$

$$\mathcal{E}_{max} = \sqrt{\frac{2qN_D(\Psi_o + V_R)}{\epsilon}} \tag{1.25}$$

The width of the depletion region is an important parameter for the calculation of junction capacitance and the "punch through" breakdown voltage. The maximum electric field determines the avalanche breakdown voltage.

1.3.1   Breakdown Voltage

When the maximum electric field, $\mathcal{E}_{max}$, exceeds the critical field of about $5x10^5$ V/cm, free electrons in the depletion region gain enough energy from the field so that when they strike a silicon atom, it ionizes producing an additional electron hole pair. This is an avalanche effect, where each conduction electron is multiplied with each impact with the silicon lattice. All resulting carriers contribute to the current. This is avalanche breakdown. The reverse voltage equals the breakdown voltage when the maximum electric field equals the critical field. Therefore, using Equation 1.25, the breakdown voltage is

$$V_{BD} = \frac{\epsilon \mathcal{E}_c^2}{2qN_D} \qquad (1.26)$$

where $\mathcal{E}_c$ is the critical field and V_{BD} is the reverse breakdown voltage applied to the junction. The built-in potential Ψ_o has been dropped. It is typically about 0.8 V. The critical field is a function of processing. It increases with doping.

   If the width of the depletion region exceeds the dimensions of the device, punch through breakdown occurs. The depletion region extends to the contact where carriers are available to contribute to current. For example, in the single-sided p^+n junction, the depletion region is mainly in the lightly doped n-side. In the depletion region, the electric field acts to keep electrons in the n-region and holes in the p-region. Any holes in the depletion region are accelerated toward the p-side by the electric field. When the depletion region reaches the contact, holes at the contact are accelerated across the depletion region toward the p-side of the junction. A large current flows. This is punch through breakdown.

Sample Problem

   A pn junction fabricated in silicon has doping densities $N_A = 10^{15}$ atoms per cm^3 and $N_D = 10^{16}$ atoms per cm^3. Calculate the built-in potential, the junction depths in both regions, and the maximum electric field with $V_R = 10$ V. Calculate the depletion width assuming a single-sided junction. How much error is created using this approximation?

Answer

   a) From Equation 1.12, we have

$$\Psi_o = 26 \; mV * ln \left[\frac{10^{15}10^{16}}{(1.5x10^{10})^2} \right] = 638 \; mV$$

b) From Equation 1.22, we have, for the p-region

$$0.638 + 10 = \frac{qx_p^2 N_A^2}{2\epsilon}\left[\frac{1}{10^{15}} + \frac{1}{10^{16}}\right]$$

$$\frac{10.638}{1.1}\frac{2\epsilon}{qN_A} = x_p^2$$

$$x_p = 3.5x10^{-4}cm = 3.5\mu m$$

From Equation 1.13, we have

$$x_n = x_p\frac{N_A}{N_D} = 0.35\mu m$$

c) From Equation 1.18, we have

$$\mathcal{E}_{max} = -\frac{1.6x10^{-19}10^{15}3.5x10^{-4}}{1.04x10^{-12}} = -5.4x10^4\frac{V}{cm}$$

d) If we assume the depletion region exists entirely within the p-region, the depletion width is equal to

$$x_d = x_p = 3.5\mu m$$

e) The actual width of the depletion region is $x_d = x_p + x_n = 3.85\mu m$. The error introduced is 10% for this example; however, if the doping difference was an order of magnitude larger, say $N_D = 10^{17}$, the error would only be 1%. Since the difference in doping for most pn junctions built today is usually a factor of 100 or more, the single-sided junction is a good approximation in many cases.

1.3.2   Junction Capacitance

When the voltage applied to the junction changes, the width of the depletion region changes. This requires charges to be added or removed. For an increase in the reverse applied voltage, the n-side is made more positive than the p-side. Electrons are removed from the n-side and holes are removed from the p-side. A positive current flows into the n-side contact and out the p-side contact. The width of the depletion region increases. The incremental capacitance is defined as the charge that flows divided by the change in voltage. The structure acts like a parallel plate capacitor with the capacitance equal to

$$C_J = \frac{\epsilon A}{x_d} \tag{1.27}$$

where A is the cross-sectional area of the junction. Since x_d, the width of the depletion region, is a function of voltage, the junction capacitance is also a function of voltage. Plugging Equation 1.24 into Equation 1.27

$$C_J = \frac{C_{J0}}{\sqrt{1 + \frac{V_R}{\Psi_o}}} \tag{1.28}$$

where

$$C_{J0} = A\sqrt{\frac{\epsilon q N_D}{2\Psi_o}} \tag{1.29}$$

Equations 1.29 and 1.27 apply to the single-sided junction with uniform doping in the p-sides and n-sides. If the doping varies linearly with distance, junction capacitance varies inversely as the cube root of applied voltage.

1.3.3   The Law of the Junction

The law of the junction is used to calculate electron and hole densities in pn junctions. It is based on Boltzmann statistics. Consider two sets of energy states. They are identical, except that set 1, at energy level E_1, is occupied by N_1 electrons and set 2, at energy level E_2, is occupied by N_2 electrons. The Boltzmann assumption is that

$$\frac{N_2}{N_1} = e^{-\frac{E_2 - E_1}{KT}} \tag{1.30}$$

In a pn junction, the built-in potential Ψ_o, across the junction causes an energy difference. The conduction band edge on the p-side of the junction is at a higher energy than the conduction band on the n-side of the junction. On the n-side of the junction, outside the depletion region, the density of electrons is N_D, the donor concentration. On the p-side of the junction, outside the depletion region, the density of electrons in the conduction band is n_i^2/N_A. Conduction band states in the n-side are occupied but conduction band states in the p-side tend to be unoccupied. Boltzmann's Equation 1.30 can be used to find the relationship between the densities of conduction electrons on the n-sides and p-sides of the junction and the junction built-in potential. Let N_1 equal the density of conduction electrons on the p-side of the junction and N_2 equal the density of electrons on the n-side of the junction. Then using Equation 1.30,

$$\frac{N_2}{N_1} = \frac{n_i^2}{N_A N_D} = e^{\frac{\Psi_o}{V_T}}$$

$$\Psi_o = V_T ln\left[\frac{n_i^2}{N_A N_D}\right]$$

where $V_T = KT/q$ is the thermal voltage.

And since potential (voltage) is energy per unit charge and the charge involved is -q, the charge of an electron, Ψ_o, the potential of the n-side of the junction relative to the p-side due to the different doping on the p-sides and n-sides: $\Psi_o = -(E_2 - E_1)/q$.

The relationship between voltage and electron energy is a point of confusion. The voltage is the negative of the energy expressed in electron volts. If electron energy is expressed in Joules, the voltage is the energy per unit charge, $V = -E/q$, where the electronic charge is $-q$. The minus sign is due to the negative charge on electrons. Where voltage is higher, electronic energy is lower. Electrons move to higher voltages where their energy is lower.

If a forward voltage is applied to the junction, it subtracts from the built-in potential. It reduces the barrier to the flow of carriers across the junction. Holes move from the p-side to the n-side and electrons move from the n-side to the p-side. This is the injection process described by the law of the junction. Boltzmann statistics predicts $p_n(0)$, the hole density at the edge of the depletion region in the n-side of the junction

$$p_n(0) = p_{n0} e^{\frac{V_a}{V_T}} \tag{1.31}$$

where $p_{n0} = n_i^2/N_D$ is the equilibrium hole concentration in the n-side and V_a is the applied voltage. Applying a forward voltage decreases the energy of the levels on the n-side occupied by holes. Equation 1.31 uses Boltzmann's statistics to determine the density of holes on the n-side of the junction as a function of the applied forward voltage V_a. With no applied forward voltage the hole density on the n-side is equal to the equilibrium density p_{n0}. With an applied forward voltage, the hole energy levels on the n-side decrease and the number of holes increase exponentially.

Equation 1.31 is referred to as the law of the junction. A similar equation applies to electrons injected into the p-side.

1.3.4 Diffusion Capacitance

Forward current in a pn junction is due to diffusion and requires a gradient of minority carriers. For example, in the p^+n single-sided junction, current is dominated by holes injected into the n-side. These holes injected into the n-region are called excess holes because they cause the number of holes to exceed the equilibrium number. The excess holes represent charge stored in the junction. If the voltage applied to the diode V_{be} changes, the number of holes stored in the n-region changes. Figure 1.7 shows a plot of the holes in the n-region as a function of x.

The number of holes in the n-region decreases from the injected value at the boundary of the n-region and the depletion region ($x = 0$) to the equilibrium hole concentration at the contact. The total charge due to the holes stored in the n-region is the total number of holes in the n-region multiplied by q, the charge per hole

$$Q = AqW_B \frac{[p_n(0) - p_{n0}]}{2} = \frac{AqW_B n_i^2}{2N_D} \left[e^{\frac{V_{be}}{V_T}} + 1 \right] \tag{1.32}$$

where $p_{n0} = n_i^2/N_D$ has been used, A is the junction area, and W_B is the distance of the n-side contact from the junction. Diffusion capacitance describes the incremental change in charge Q due to an incremental change in voltage V_{be}. For V_{be} greater than a few V_T, $e^{\frac{V_{be}}{V_T}} \gg 1$ and the 1 can be dropped in Equation 1.32. Then the diffusion capacitance is

$$C_{diff} = \frac{\partial Q}{\partial V_{be}} = \frac{AqW_B n_i^2}{2N_D V_T} e^{\frac{V_{be}}{V_T}} \tag{1.33}$$

Diffusion capacitance is significant only in forward biased pn junction diodes where it increases exponentially with applied voltage.

1.4   Diode Current

Diffusion is the dominant mechanism for current flow in pn junctions. Carriers injected across the depletion region produce a carrier density gradient that results in diffusion current flow. Holes are injected from the p-side to the n-side and electrons are injected from the n-side to the p-side. Current density due to diffusion is a function of the concentration gradient and of the carrier mobility. Consider the component of current due to holes injected into the n-region. Current density (amperes per cm^2) is

$$J_p = -qD_p \frac{dp}{dx} \tag{1.34}$$

where D_p is the diffusion constant in cm^2 per second, q is electronic charge in coulombs, and $\frac{dp}{dx}$ is the hole concentration gradient in holes per cm^3 per cm (cm^{-4}).

In the short diode approximation, the width of the n neutral region from the depletion region to the contact W_B is short, recombination is neglected. This is true for most bipolar integrated devices where dimensions are less than a few microns. When recombination is neglected, the hole density gradient is constant as shown in Figure 1.7.

The hole concentration gradient is the slope of $p_n(x)$ as shown in Figure 1.7:

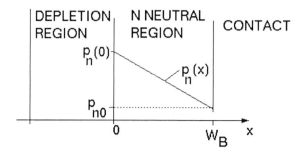

Figure 1.7   Holes injected into the n-side of the pn junction become minority carriers that diffuse across the n neutral region. $P_{n0} = n_i^2/N_D$ is the equilibrium density of holes in the n-region.

$$\frac{dp}{dx} = -\frac{p_n(0) - p_{n0}}{W_B} \tag{1.35}$$

Heavy doping at the contact reduces carrier lifetime and causes the hole concentration to equal the equilibrium concentration, p_{n0}. Using the law of the junction, Equation 1.31, and Equation 1.35, the hole current density, Equation 1.34 becomes

$$J_p = \frac{qD_p p_{n0}}{W_B}\left[e^{\frac{V_{be}}{V_T}} - 1\right] \tag{1.36}$$

where $P_{n0} = n_i^2/N_D$.

  There is a similar expression for the current due to electrons injected in to the p-side. The total current density is the sum of the electron and hole components

$$J = \left[\frac{qD_p n_i^2}{N_D W_B} + \frac{qD_n n_i^2}{N_A W_A}\right]\left[e^{\frac{V_{be}}{V_T}} - 1\right] \tag{1.37}$$

where W_A is the distance of the contact on the p-side to the depletion region. Typically one side of the junction is more heavily doped than the other. For the case where the p-side is the heavily doped side, hole current dominates over electron current and Equation 1.37 reduces to

$$J = \frac{qD_p n_i^2}{N_D W_B}\left[e^{\frac{V_{be}}{V_T}} - 1\right] \tag{1.38}$$

The diode current in amperes is the current density multiplied by the cross-sectional area A

$$I = \frac{AqD_p n_i^2}{N_D W_B}\left[e^{\frac{V_{be}}{V_T}} - 1\right] \tag{1.39}$$

We now define a process constant called saturation current I_s where

$$I_s = \frac{qD_pAn_i^2}{N_DW_B} \tag{1.40}$$

Equation 1.39 becomes

$$I = I_s \left[e^{\frac{V_{be}}{V_T}} - 1 \right] \tag{1.41}$$

Equation 1.41 is called the rectifier equation. It describes the pn junction voltage current relationship. It is the governing equation not only for pn junction diodes but bipolar transistors as well. For typical integrated circuit diodes and transistors I_s is quite small (10^{-16} is a typical value). Since I_s is small, the term in the brackets has to be large for measurable currents. That means the "1" in the bracket is negligible and can be dropped for V_{be} more than a few V_T. For $V_{be} = 0.1\ V$, $e^{\frac{V_{be}}{V_T}} = 46.8$, since $V_T = 0.026\ V$ at room temperature. Equation 1.41 becomes

$$I = I_s e^{\frac{V_{be}}{V_T}} \tag{1.42}$$

Small changes in V_{be} produce large changes in current. For typical values of I_s, V_{be} is about $0.7\ V$ for forward conducting silicon diodes.

Example

If $V_{be} = 0.7\ V$ when $I = 100\ \mu A$, what is I_s?

Answer

$$I_s = I e^{-\frac{V_{be}}{V_T}} = 10^{-4} e^{-\frac{0.7}{0.026}} = 2x10^{-16}\ A$$

1.5 Bipolar Transistors

The structure of a vertical npn transistor is shown in Figure 1.8. The transistor is formed by growing a lightly doped n-type epitaxial layer on a p-type substrate. This layer becomes the collector. The p-type base is diffused into the epitaxial collector and the n-type emitter is diffused into the base as shown in Figure 1.8. A p-type isolation well (ISO) is diffused from the surface to the substrate. During circuit operation, the substrate is biased at the lowest voltage in the circuit. This reverse biases the collector-iso pn junction isolating the collector epi. In normal operation the base-emitter pn junction is forward biased and the base-collector pn junction is reversed biased. Since the emitter is more

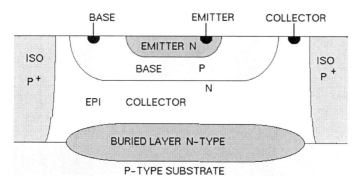

Figure 1.8   The structure of a vertical npn transistor is shown. The p-type substrate and iso are held at a low voltage, reverse biasing the substrate-epi pn junction to isolate the transistor. The high conductivity buried layer provides a low resistance path for collector current.

heavily doped than the base, the forward current across the base-emitter junction is dominated by electrons. The electrons injected into the base cause an electron concentration gradient in the base that results in diffusion of electrons across the p-type base.

1.5.1   Collector Current

The law of the junction, Equation 1.31, expresses the electron concentration in the base at the edge of the base-emitter depletion region, as a function of the voltage applied to the base-emitter junction. It also expresses the electron concentration in the base at the edge of the base-collector depletion region as a function of the voltage applied to the base-collector junction. In the base at the edge of the base-emitter depletion region, the electron concentration is

$$n_p(0) = \frac{n_i^2}{N_D} e^{-\frac{V_{be}}{V_T}} \qquad (1.43)$$

The electron concentration in the base at the emitter is many orders of magnitude greater than the equilibrium concentration. In the base at the collector the electron concentration is

$$n_p(W_B) = \frac{n_i^2}{N_D} e^{-\frac{V_{bc}}{V_T}} \qquad (1.44)$$

where V_{bc} is the voltage applied to the base relative to the collector. In normal operation the collector is biased positive relative to the base, so V_{bc} is a negative voltage. The exponent in Equation 1.44 is a large negative number and the electron concentration in the base at the collector approaches zero. This is illustrated in Figure 1.9.

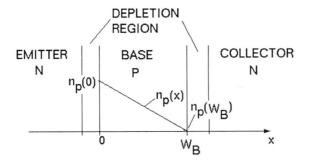

Figure 1.9 The gradient of the minority carrier concentration $\frac{dn_p(x)}{dx}$ in the base determines the collector current.

Electrons diffusing across the base to the collector results in collector current that depends on the electron density gradient in the base

$$I_c = -A_E q D_n \frac{dn}{dx} \tag{1.45}$$

where A_E is the emitter area. The minus sign is because I_c flows in the negative x direction.

For a transistor biased in the normal operating range, V_{bc} is a negative number and $n_p(W_B)$ approaches zero. From Figure 1.9

$$\frac{dn}{dx} = -\frac{n_p(0)}{W_B} \tag{1.46}$$

Using Equation 1.46 in Equation 1.45,

$$I_c = I_s e^{\frac{V_{be}}{V_T}} \tag{1.47}$$

where

$$I_s = \frac{A_E q D_n n_i^2}{W_B N_D} \tag{1.48}$$

and where N_D is the base doping, donors per cm^3.

Equation 1.47 describes the collector current as a function of base to emitter voltage. It is an important equation, widely used in bipolar circuit design.

1.5.2 Base Current

Bipolar transistors are current gain devices. The collector current is a multiple of the base current. The current gain $\beta = I_c/I_b$ varies over a wide range for transistors produced by a given process. Generally better, higher gains are achieved by reducing base current I_b. Two physical mechanisms are responsible for base current. The first is due to holes injected from the base to the emitter. With the base-emitter junction forward biased, electrons are injected from the emitter to the base and holes are injected from the base to the emitter. The electrons diffuse across the base to the collector where they form the main component of collector current. Holes injected into the emitter from the base are the main source of base current. Every hole leaving the base has to be replaced by a hole from the base contact, thereby producing base current. Holes are injected from the base to the emitter in order to maintain the hole density $p_n(0)$ in the n-type emitter at the edge of the base-emitter depletion region, predicted by the law of the junction

$$p_n(0) = p_{no} e^{\frac{V_{be}}{V_T}} \tag{1.49}$$

where $p_{no} = n_i^2/N_{DE}$ is the equilibrium hole concentration in the emitter. N_{DE} is the donor doping concentration in the emitter.

Holes injected into the emitter diffuse to the emitter contact. Assuming negligible recombination in the emitter, this hole current is given by Equation 1.41 applied here to hole current in the npn base-emitter junction

$$I_b = I_{se} \left[e^{\frac{V_{be}}{V_T}} - 1 \right] \tag{1.50}$$

where

$$I_{se} = \frac{q D_p A_E n_i^2}{N_{DE} W_E} \tag{1.51}$$

where D_p is the diffusion constant for holes in the emitter and W_E is the distance of the emitter-base junction to the emitter contact.

Recombination in the base also contributes to base current. Every hole that recombines with an electron has to be replaced by a hole from the base contact. This contributes to base current. For modern integrated circuit transistors, this component is small. Here we ignore it.

The transistor gain β is the ratio of I_c/I_b. Using Equations 1.47 and 1.50

$$\beta = \frac{I_c}{I_b} = \frac{D_n}{D_p} \frac{W_E}{W_B} \frac{N_{DE}}{N_A}. \tag{1.52}$$

High β is achieved by keeping the width of the base W_B small and doping the emitter more heavily than the base.

1.5.3   Ebers-Moll Model

The Ebers-Moll model describes the large signal DC operation of the bipolar transistor. Consider the distribution of minority carriers shown in Figure 1.10. We are interested in three components of current:

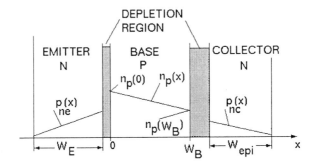

Figure 1.10   Minority carrier distribution in an npn transistor.

1. I_{pe} holes flowing in the n-type emitter.

2. I_{nc} electrons flowing in the p-type base.

3. I_{pc} holes flowing in the n-type collector.

I_{nc} is composed of electrons injected from the emitter that diffuse across the base and are swept into the collector by the base-collector junction potential. The emitter current is composed of this current plus holes diffusing across the emitter

$$I_E = -(I_{pe} + I_{nc}) \tag{1.53}$$

The collector current is due to electrons diffusing across the base to the base-collector depletion region, and holes diffusing across the collector to the base-collector depletion region

$$I_C = I_{nc} - I_{pc} \tag{1.54}$$

Here we observe the convention of positive currents flowing into the transistor. The current flow mechanism is diffusion

$$I_{nc} = A_E q D_n \frac{dn}{dx} = A_E q D_n \frac{n_p(0) - n_p(W_B)}{W_B} \tag{1.55}$$

Invoking the Law of the Junction, Equation 1.31, to determine carrier densities

$$I_{nc} = \frac{A_E q D_n n_i^2}{W_B N_A} \left[e^{\frac{V_{be}}{V_T}} - e^{\frac{V_{bc}}{V_T}} \right] \tag{1.56}$$

Similarly,

$$I_{pe} = \frac{A_E q D_{pe} n_i^2}{W_E N_{de}} \left[e^{\frac{V_{be}}{V_T}} - 1 \right] \tag{1.57}$$

and

$$I_{pc} = \frac{A_C q D_{pc} n_i^2}{W_{epi} N_{dc}} \left[e^{\frac{V_{bc}}{V_T}} - 1 \right] \tag{1.58}$$

where A_E is the emitter area, q is the electronic charge, D_n is the electron diffusion constant in the base, n_i is the intrinsic carrier concentration, W_B is the base width, N_A is the base doping, $V_T = KT/q$ is the thermal voltage, D_{ne} is the diffusion constant in the emitter, W_E is the emitter width, N_{de} is the emitter doping, A_C is the area of the collector-base junction, D_{pc} is the hole diffusion constant in the collector, W_{epi} is the width of the collector, and N_{dc} is the collector doping.

Rewriting Equations 1.56, 1.57, and 1.58 using constants, A, B, C, where

$$A = \frac{A_E q D_n n_i^2}{W_B N_A}$$

$$B = \frac{A_E q D_{pe} n_i^2}{W_E N_{de}}$$

$$C = \frac{A_C q D_{pc} n_i^2}{W_{epi} N_{dc}}$$

Using the constants A, B, and C in Equations 1.56, 1.57, and 1.58:

$$I_{nc} = A \left[e^{\frac{V_{be}}{V_T}} - e^{\frac{V_{bc}}{V_T}} \right]$$

$$I_{pe} = B \left[e^{\frac{V_{be}}{V_T}} - 1 \right] \qquad (1.59)$$

$$I_{pc} = C \left[e^{\frac{V_{bc}}{V_T}} - 1 \right]$$

Plugging Equations 1.59 into Equations 1.53 and 1.54:

$$I_E = A \left[e^{\frac{V_{be}}{V_T}} - e^{\frac{V_{bc}}{V_T}} \right] + B \left[e^{V_{be}} - 1 \right]$$

$$I_E = -A \left[e^{\frac{V_{be}}{V_T}} - e^{\frac{V_{bc}}{V_T}} \right] + C \left[e^{V_{bc}} - 1 \right]$$

Note there are only three constants A, B, and C.

If the following new constants are defined:

$$I_{ES} = -(A + B)$$

$$I_{CS} = -(C - A)$$

$$\alpha_R I_{CS} = \alpha_F I_{ES} = -A$$

then

$$I_E = -I_{ES}(e^{\frac{V_{be}}{V_T}} - 1) + \alpha_R I_{CS}(e^{\frac{V_{bc}}{V_T}} - 1) \qquad (1.60)$$

$$I_C = \alpha_F I_{ES}(e^{\frac{V_{be}}{V_T}} - 1) - I_{CS}(e^{\frac{V_{bc}}{V_T}} - 1) \qquad (1.61)$$

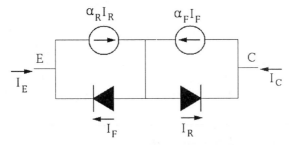

Figure 1.11   Ebers-Moll model $I_F = I_{ES}(e^{\frac{V_{be}}{V_T}} - 1)$ $I_R = I_{CS}(e^{\frac{V_{bc}}{V_T}} - 1)$.

Equations 1.60 and 1.61 describe the Ebers-Moll model. A schematic diagram for the Ebers-Moll model, is shown in Figure 1.11. In the normal operating range, the base-collector junction is reversed biased. V_{bc} is a negative voltage.

$$e^{\frac{V_{bc}}{V_T}} \Longrightarrow 0$$

Under this condition Equations 1.60 and 1.61 become

$$I_E = -I_{ES}(e^{\frac{V_{be}}{V_T}} - 1) - \alpha_R I_{CS} \qquad (1.62)$$

$$I_C = \alpha_F I_{ES}(e^{\frac{V_{be}}{V_T}} - 1) + I_{CS} \qquad (1.63)$$

Neglecting the small leakage current I_{CS}

$$I_E = -I_{ES}(e^{\frac{V_{be}}{V_T}} - 1) \qquad (1.64)$$

$$I_C = -\alpha_F I_E \qquad (1.65)$$

α_F is slightly less than one. The base current is

$$I_B = -(I_C + I_E) = I_C(\frac{1}{\alpha_F} - 1) \qquad (1.66)$$

The transistor current gain is

$$\frac{I_C}{I_B} = h_{FE} = \beta_F = \frac{\alpha_F}{1 - \alpha_F} \qquad (1.67)$$

When $\beta_F = 100$, $\alpha_F = 0.99$. For larger β, α gets closer to 1.

1.5.4   Breakdown

When the electric field in a reversed biased pn junction exceeds a critical value of about $3x10^5$ V/cm the junction breaks down causing current to flow. In breakdown, the junction voltage is stable over a wide range of currents. A pn junction in breakdown is used as a voltage reference called a "zener diode." If current is limited, the junction recovers when the reverse voltage is reduced. Designers use these zeners for a wide variety of clipping and protection circuits. Transistors are designed to operate over a range of voltages without breakdown occuring. In bipolar transistors, higher breakdown voltages are achieved by reducing collector (epi) doping.

In the normal operating mode, breakdown in bipolar transistors occurs at the reversed biased base-collector junction. There are two breakdown voltages of interest: BV_{CBO} and BV_{CEO}. BV_{CBO} is less than BV_{CEO}. BV_{CEO} is the collector-base breakdown voltage with the emitter open. BV_{CBO} is the collector-emitter breakdown voltage with the base open.

Electron-hole pairs are generated at the base-collector junction by the breakdown process. The collector-base junction electric field moves the holes into the p-type base. This constitutes base current and is amplified by transistor action producing a larger collector current. Holes accumulating in the floating base raise the base potential. This forward biases the base-emitter junction, turning the transistor on. Assuming an avalanche multiplication mechanism, we can derive a relationship between BV_{CBO} and BV_{CEO}. As the collector-base voltage V_{cb} approaches the breakdown voltage BV_{CBO} currents normally flowing through the junction are multiplied by a factor M given by the empirical relation

$$M = \frac{1}{1 - \left(\frac{V_{cb}}{BV_{CBO}}\right)^n} \tag{1.68}$$

Since the avalanche multiplication process increases the collector current by a factor of M

$$I_C = -M\alpha_F I_E$$

$$h_{FE} = \frac{I_C}{I_B} = \frac{M\alpha_F}{1 - M\alpha_F}$$

At breakdown, $M = 1/\alpha_F$ and the current gain h_{FE} goes to infinity. Setting M equal to $1/\alpha_F$ and V_{cb} equal to BV_{CEO} in Equation 1.68

$$BV_{CEO} = BV_{CBO} \sqrt[n]{1 - \alpha_F} \approx BV_{CBO}(h_{FE})^{-\frac{1}{n}} \tag{1.69}$$

BV_{CEO} can be substantially less than BV_{CBO}. n is between 2 and 4 in silicon. If $h_{FE} = 100$ and $n = 3$, BV_{CEO} is approximately one fifth of BV_{CBO}.

1.6   MOS Transistors

A representation of a MOS transistor is shown in Figure 1.12. The gate-oxide-substrate form the metal-oxide-silicon (MOS) structure. The

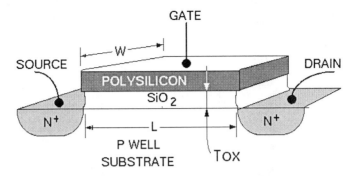

Figure 1.12 NMOS Transistor.

aluminum gates of early transistors have been replaced by polycrystalline silicon (POLY) because poly has a higher melting point. This permits the gate to be placed before the source and drain. With the gate in place first, it acts as a mask for the source and drain diffusions, producing self-aligned structures. The heavily doped poly has a high conductivity. It behaves like a metal.

Current flow between the source and the drain is controlled by the gate voltage. For the NMOS transistor shown, a positive gate voltage attracts electrons to the p-type substrate region between the source and drain, turning the transistor on. When the voltage applied to the gate is below a threshold, there are no mobile electrons in the channel between the source and drain. No current flows. The drain to substrate and substrate to source silicon regions represent two back to back pn junctions, blocking current flow in either direction. With a positive voltage applied to the drain relative to the source, the drain-substrate pn junction is reversed biased. The source substrate pn junction is forward biased. A positive gate voltage attracts mobile electrons to the interface between the silicon and the oxide below the gate. These electrons form the channel. Channel electrons drifting to the substrate-drain pn junction are swept across by the drain-substrate junction voltage. This forms the drain current.

For a channel of mobile electrons to form, the gate to source voltage must exceed a threshold voltage. The MOS structure is a capacitor formed by the poly gate, the oxide, and the silicon substrate. A positive voltage on the gate relative to the substrate results in a positive charge on the poly and a negative charge in the substrate at the substrate-oxide interface. Initially, at low gate voltages, the negative charge in the p-type silicon substrate is due to the absence of positively charged holes. This negative charge is ionized acceptor atoms. As the gate voltage becomes more positive, a depletion region forms as holes are repelled by the positive gate voltage. As the gate voltage increases further, the negative

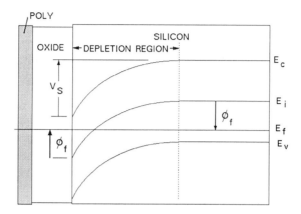

Figure 1.13   Band bending at the onset of moderate inversion.

charge in the silicon increases to include electrons as well as ionized acceptors. The electrons are mobile and can contribute to current flow.

A positive gate voltage reduces electron energy in the silicon under the gate. This can be represented using the band diagram shown in Figure 1.13. With electrons as carriers in the p-type silicon, the channel is said to be inverted. It is convenient to define the onset of moderate inversion to be when the bands at the silicon surface at the oxide interface are $2\phi_f$ below their values in the bulk away from the surface. The surface is at a voltage $2\phi_f$ above the bulk due to the influence of the gate. Recall that voltage is energy per unit charge. Since electrons have a negative charge, when electron energy decreases, voltage increases. Also ϕ_f, the Fermi energy, is the position of the intrinsic energy level relative to the Fermi level in the bulk semiconductor as shown in Figure 1.13.

The gate to bulk voltage at the onset of moderate inversion is the sum of:

1. The surface potential V_s. This is the voltage at the oxide interface relative to the bulk.

2. The voltage across the oxide.

3. The contact potential between the gate and the bulk Φ_{ms}.

$$V_{GB} = V_s + V_{ox} + \Phi_{ms} \qquad (1.70)$$

At the onset of moderate inversion $V_s = 2\phi_f$ as shown in Figure 1.13. The voltage across the oxide is the electric field in the oxide multiplied by the oxide thickness t_{ox}. From Gauss' law, the electric field in the oxide is the charge per unit area on the gate divided by the oxide permittivity: $\mathcal{E}_{ox} = Q_G/\epsilon_{ox}$. The voltage across the oxide is

$$V_{ox} = \mathcal{E}_{ox}t_{ox} = \frac{Q_G}{\epsilon_{ox}}t_{ox} \qquad (1.71)$$

Since the positive charge on the gate must be balanced by negative charge in the silicon and in the oxide

$$Q_G = Q_B - Q_{ox} + Q_I \qquad (1.72)$$

where Q_B is the charge due to ionized acceptors in the depletion region. $Q_B = qN_A x_d$ where N_A is the substrate doping and x_d is the width of the depletion region. Q_{ox} is positive charge trapped in the oxide. Here we assume Q_{ox} is all trapped at the oxide silicon interface. Q_I is charge due to mobile electrons in the channel. At the onset of moderate inversion, Q_I is small and does not contribute to Q_G. The charge Q_B, due to ionized acceptors in the depletion region depends on V_s, the surface potential. V_s is the amount the bands are bent. V_s is the voltage across the depletion region. Equation 1.24 describing the depletion region in a pn junction can be used to determine the width of the depletion region and the charge Q_B

$$Q_B = \sqrt{2qN_A\epsilon V_s}$$

At the onset of moderate inversion $V_s = 2\phi_f$

$$Q_B = \sqrt{4qN_A\epsilon\phi_f}$$

From Equations 1.70, 1.71 and 1.72

$$V_{GB} = \Phi_{ms} + V_s + \frac{Q_B - Q_{ox}}{\epsilon_{ox}} t_{ox} \qquad (1.73)$$

Since the gate capacitance per unit area is

$$C_{ox} = \frac{t_{ox}}{\epsilon_{ox}}$$

$$V_{GB} = \Phi_{ms} + V_s + \frac{Q_B - Q_{ox}}{C_{ox}}$$

At the onset of moderate inversion $V_s = 2\phi_f$.

$$V_{GB} = \Phi_{ms} - \frac{Q_{ox}}{C_{ox}} + 2\phi_f + \frac{\sqrt{4qN_A\epsilon\phi_f}}{C_{ox}} \qquad (1.74)$$

V_{GB}, given in Equation 1.74, is the gate to bulk voltage at the threshold, when the transistor begins to turn on. When the bulk is connected to the source V_{GB}, the gate to bulk voltage at the onset of moderate inversion is V_{TO}, the gate to source threshold voltage at zero bulk bias

$$V_{TO} = \Phi_{ms} - \frac{Q_{ox}}{C_{ox}} + 2\phi_f + \gamma\sqrt{2\phi_f} \qquad (1.75)$$

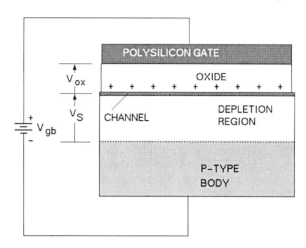

Figure 1.14 The gate to body voltage, V_{GB} is the sum of the surface potential, V_s, the voltage across the oxide, Vox, and the body to gate contact potential Φ_{ms}.

where $\gamma = \sqrt{2qN_A\epsilon}/C_{ox}$. γ (GAMMA) is the body effect parameter.

The contact potential between the gate and the bulk Φ_{ms} contributes to the gate voltage. Consider Figure 1.14. When the gate is shorted to the bulk, $V_{GB} = 0$, there is an internal contact potential that can be expressed in terms of the positions of the Fermi levels relative to the intrinsic level in the polysilicon gate and the bulk. In the bulk, the position of the Fermi level relative to the intrinsic level is ϕ_f. In the gate, the position of the Fermi level depends on the gate material. The two cases of interest for MOS transistors are polysilicon gates heavily doped either n-type or p-type. For n-type poly gates the Fermi level approaches the conduction band and is $E_g/2$ above the intrinsic level. For p-type gates the Fermi level approaches the valence band and is $E_g/2$ below the intrinsic level. When the gate is shorted to the bulk, charge moves and the energy bands adjust so the Fermi levels will be the same in both materials. This results in a contact potential of

$$\Phi_{ms} = \pm\frac{E_g}{2} - \phi_f \tag{1.76}$$

where $E_g/2$ is positive for p-type poly gates and negative for n-type poly gates. When the gate is a metal instead of polysilicon, this contact potential would be expressed as the difference in the work functions of the gate and bulk.

In the complementary metal-oxide-semiconductor (CMOS) structure shown in Figure 1.15, NMOS and PMOS transistors work together to realize circuit functions. Figure 1.15 shows one CMOS implementation with an NMOS transistor in a p-well and a PMOS transistor in the n-type epitaxial layer. While most of the discussion in this chapter involves the NMOS transistor, the PNOS transistor functions in the same way with the difference that diffusion types are reversed. N-type is replaced by p-type, and p-type is replaced by n-type. Voltage polarities and current directions are also reversed. Current flow in the channel of PMOS transistors is due to holes rather than electrons. As more holes are attracted to the channel, the more negative the gate to source voltage becomes. This complementary nature of NMOS and PMOS transistors is useful in the design of analog and digital circuits.

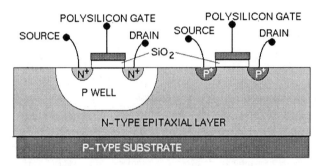

Figure 1.15 CMOS structure.

1.6.1 Simple MOS Model

A simple model for the MOS transistor, useful for hand calculations, can be derived by considering the channel to be a variable resistor whose value depends on the gate to channel voltage, then summing the voltage across this channel resistance from the source to the drain.

Here we use the source as the voltage reference point by setting the source voltage equal to zero. With the source as the voltage reference, $V_g = V_{gs}$. The drain current I_D flowing in the channel causes the channel voltage V_{cs}, and therefore the gate to channel voltage V_{gc} to be a function of the distance x from the source as shown in Figure 1.16. V_{gc} is the voltage across the oxide. The channel consists of electrons attracted by the positive gate voltage. The mobile charge in the channel is

$$Q(x) = C_{ox}(V_{gc} - V_{th}) \text{ Coul/square meter} \qquad (1.77)$$

where C_{ox} is the capacitance of the gate oxide per unit gate area. The channel does not exist until V_{gc} is greater than the threshold voltage V_{th}. That is, $V_{gs} - V_{cs} - V_{th} > 0$. Also the maximum value of V_{cs} is

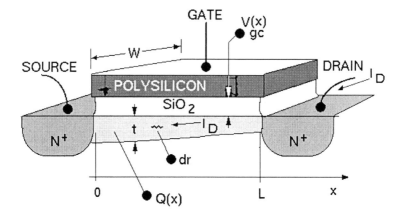

Figure 1.16   The channel resistance varies with x because channel voltage and therefore mobile charge varies with x.

V_{ds}. Therefore, since the largest value of V_{cs} is V_{ds}, $V_{gs} - V_{th}$ must be greater than V_{ds} for this derivation to hold. Otherwise, at the drain end of the channel where the channel voltage is the greatest, there will be no mobile charge. The resistive channel can be represented as a series of small resistances dr. The current I_D flowing through these resistances causes the voltage drop in the channel. The voltage across each of these incremental resistances is $dV_{cs} = I_D dr$ where $dr = dx/(\sigma t W)$ where W is the width of the gate and σ is the conductivity of the channel and t is the effective channel thickness.

$$\sigma = \text{charge per unit volume times the mobility}$$

$$\sigma = \frac{\mu Q(x)}{t}$$

where $Q(x)$ is the mobile channel charge per unit gate area. $Q(x)/t$ is the mobile channel charge per unit volume. Therefore, $dr = dx/[\mu Q(x)W]$. Using Equation 1.77 for $Q(x)$

$$dV_{cs} = \frac{I_D dx}{\mu W C_{ox}(V_{gs} - V_{cs} - V_{th})}$$

Rearrange and integrate

$$\int_0^L I_D \, dx = \int_0^{V_{ds}} \mu W C_{ox}(V_{gs} - V_{th} - V_{cs}) \, dV_{cs}$$

$$I_D = \mu C_{ox}(W/L)\left[(V_{gs} - V_{th})V_{ds} - \frac{V_{ds}^2}{2}\right] \qquad V_{ds} < V_{gs} - V_{th}$$

I_D increases as V_{ds} increases to a maximum value that occurs when $V_{ds} = V_{gs} - V_{th}$.

Saturation

Drain current is self limiting. As the drain to source voltage V_{ds} increases, drain current increases. This increases the channel voltage reducing the gate to channel voltage and the mobile channel charge. When $V_{ds} > V_{gs} - V_t$, $Q(x)$ vanishes at the drain end of the channel causing the transistor to operate in the saturation or constant current region. Voltages applied to the drain are absorbed across the channel-drain depletion region where no mobile charge exists. The drain is more positive than the channel. Channel electrons entering this depletion region are swept into the drain by the built-in potential and the voltage applied to the drain. Since increases in drain voltages appear across the drain-channel depletion region, channel voltages and therefore channel current does not change with drain voltage. The drain current remains constant with changes in drain voltage.

With all voltages referenced to the source, V_g becomes V_{gs} and the drain current is

$$
I_D = \begin{cases} \mu_n C_{ox} \left[(V_{gs} - V_{th}) V_{ds} - \frac{V_{ds}^2}{2} \right] & V_{ds} \leq V_{gs} - V_{th} \\[2ex] \frac{\mu_n C_{ox}}{2} (V_{gs} - V_{th})^2 & V_{ds} \geq V_{gs} - V_{th} \end{cases} \tag{1.78}
$$

Equation 1.78 is a simple model useful for hand calculations.

1.7 DMOS Transistors

Double diffused MOS (DMOS) transistors rely on the control of the lateral diffusion to achieve short channel lengths. One implementation is shown in Figure 1.17. Polysilicon is grown over a thin oxide and a small hole is etched in the polysilicon. A p-well is diffused through the hole into the n-type epitaxial layer. The p-well diffuses laterally as well as vertically into the epi. The center of the hole is masked and a second diffusion is done. This time the n-type source is diffused through the hole. These two diffusions define the MOS transistor. The epi acts as the drain. The channel forms in the p-well between the source and the epi drain. The length of the channel is the difference between the lateral diffusion of the p-well and the source. The width is approximately the perimeter of the hole in the polysilicon. A heavily doped P-diffusion is placed in the center of the device through the n-source diffusion to the p-well. This diffusion is used to make contact with the p-well. The hole is then covered with metal. A metal contact shorts the p-well to the

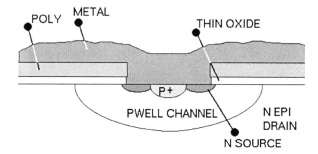

Figure 1.17 A double diffused DMOS transistor is fabricated by diffusing first a p-well into the n-type epi through a hole in the polysilicon. A second n-type diffusion forms the source. The epi acts as drain and the channel forms in the p-well. Channel length depends on lateral diffusion of the p-well and the n-type source.

source. The drain contact is made to the epi. Arrays of these devices result in efficient layouts for power transistors.

1.8 Zener Diodes

Pn junctions operating in breakdown are used as voltage references and in clipping and clamping circuits. The breakdown voltage varies inversely with the square root of the doping in the lightly doped side of the junction, as described in Equation 1.26. If a zener is to survive breakdown, it is important that the current be distributed across the cross-sectional area of the junction. Doping and electric field nonuniformity results in breakdown occurring first where the doping and the electric field are largest. The deep buried layer iso zener, shown in Figure 1.18, has a smooth cross-sectional area with a uniform distribution of dopants. Currents tend to be distributed over a larger area than in the surface, shallow-n shallow-p zener shown in Figure 1.18, where doping varies with distance from the surface and has sharp corners where the electric field is large. These zeners are destroyed by large currents. They fail at corners near the surface. If the current is limited by external circuits so that power dissipation in the junction is maintained within safe limits, the junction is not damaged. Pn junctions operated in reverse breakdown are called "zener diodes" and are used as voltage references or in clipping and clamping circuits for protection of sensitive structures.

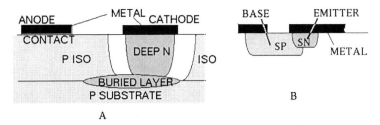

Figure 1.18 A. The deep lying pn junction formed by the buried layer and the isolation diffusion breaks down at about 12 V and can conduct large currents. B. The pn junction formed at the surface using shallow-n (SN) and shallow-p (SP) diffusions breaks down at about 6 V.

1.9 EpiFETs

Large value resistors can be achieved by enhancing the sheet resistance of the epitaxial layer with a junction field effect transistor called an *epiFET*.

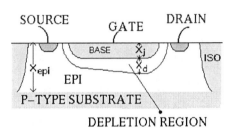

Figure 1.19 Current flow in the n-type epitaxial region is restricted by the depletion region associated with the base-epi pn junction. Sheet resistance of the epiFET can be significantly higher than the epi sheet resistance.

P-type base diffusion, shown in Figure 1.19, in the lightly doped n-type epitaxial layer(epi) forms a junction field effect transistor(JFET). Current flowing in the epi is modulated by voltages on the P-type base diffusion acting as the JFET gate. The depletion region associated with the base-epi pn junction increases in size as the gate becomes more negative relative to the epi. As the width of the depletion region increases, it penetrates further into the epi, robbing the epi of carriers. This reduces current flow in the epi creating an apparent increase in epi resistance. The width of the depletion region x_d increases when the epi to gate voltage increases. Since the drain is at a higher voltage than the source, the depletion region is wider at the drain. Current tends to be self-limiting, as shown in Figure 1.20, since it increases the epi voltage and therefore

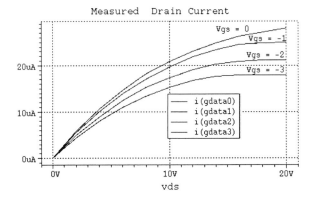

Figure 1.20 Measured epiFET characteristics are shown. The threshold voltage VTO is about -20 V. The constant current region is $Vds > Vgs - V_{th}$. For $Vgs = -3$, the constant current region starts at about $Vds = 17\ V$. [7]

the width of the depletion region. No buried layer is present because it would short out the epiFET.

In a typical configuration three regions, the gate, isolation well and the substrate are connected to the negative supply. As the epi becomes more positive, the depletion regions from all three regions extend into the epi reducing the cross-section for current flow.

1.10 Chapter Exercises

1. Calculate the resistivity of silicon doped with $5x10^{18}$ phosphorus(group 5) atoms per cubic centimeter. Assume total ionization.

2. Calculate the conductivity of intrinsic silicon.

3. Show that
$$J_n = q\mu_n \mathcal{E}$$
is equivalent to Ohm's law.

4. Calculate the position of the Fermi level relative to the intrinsic level for the silicon in problem 1.

5. A silicon wafer is doped with 10^{16} acceptors per cm^3 and then doped with $5x10^{17}$ donors per cubic cm^3. What is the resistivity? Estimate mobility using Figure 1.2.

6. Consider a silicon pn junction. The p-side is uniformly doped with 10^{18} acceptors/cm^3. The n-side is doped with 10^{16} donors/cm^3.

   • What is the built-in potential?

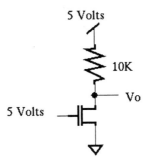

Figure 1.21 Figure for problem 9.

- What is the position of the Fermi level relative to the intrinsic level on the p-side of the junction?

7. For the pn junction in problem 6, the junction area is 10 microns by 10 microns. What is the saturation current I_s. Use mobility vs doping curves (Figure 1.2).

8. Consider a pn junction with the P-side doped with $N_A = 10^{20} cm^{-3}$. Approximately, what is the required doping on the N-side to obtain a breakdown of 20 V? Use the one-sided step junction approximation.

9. A 10 K resistor is in series with an NMOS transistor as shown in Figure 1.21: $[W/L]\mu_n Cox = 10^{-5}$. The threshold voltage is one volt. What is the output voltage, Vo?

References

[1] S. M. Sze and J. C. Irvin, *Resistivity, Mobility and Impurity Levels in GaAs, Ge, and Si at 300° K*, Solid-State Electronics, Vol 11, pp. 599-602, 1968.

[2] S. M. Sze, *Physics of Semiconductor Devices*, Wiley-Interscience, New York, 1969.

[3] Edward S. Yang, *Microelectronic Devices*, McGraw-Hill, New York, 1988.

[4] P.R. Gray and R.G. Meyer, *Analysis and Design of Analog Integrated Circuits*, 2nd edition, Wiley, New York, c. 1984, pp. 1-5.

[5] R.S. Muller and T.I. Kamins, *Device Electronics for Integrated Circuits*, 2nd edition, Wiley, New York, c. 1986, pp. 15-27, 173-188, 235-244.

[6] K. Lee, M. Shur, T.A. Fjeldy and T. Ytterdal, *Semiconductor Device Modeling for VLSI*, Prentice Hall, Englewood Cliffs, NJ, c. 1993, p. 63.

[7] Shelby Raymond, private communication, January 1999.

chapter 2

Device Models

2.1 Introduction

Models are mathematical descriptions that predict performance. They can be physical or empirical. Physical models are based on device physics and can be related to physical properties. Empirical models fit measurements to mathematical descriptions that do not necessarily correspond to device physics. Physical models are easier to adapt when parameters such as doping levels or device dimensions change.

Modeling is a tradeoff between accuracy and utility. Exact models tend to be more complex than approximate ones. The model to use is the simplist one that provides the required accuracy. Models for hand calculation, where computational power is limited, should be simple. Even with high speed computers, complex models can make the simulations of large systems prohibitive.

2.2 Bipolar Transistors

2.2.1 Early Effect

Increasing the voltage across the transistor V_{CE} results in an increase in transistor current I_C. The physical cause, is a decrease in the width of the base. As V_{CE} is increased, the reverse voltage on the collector-base pn junction increases. The collector-base depletion region extends further into the base, effectively reducing the base width. Since collector current varies inversely with base width, collector current increases. The slope of I_C vs. V_{CE} in the normal operation range is modeled by the Early voltage as shown in Figure 2.1.

2.2.2 High Level Injection

The simple model we used in Section 1.5.2 for β breaks down at high and low current levels. At high current levels, high level injection effects

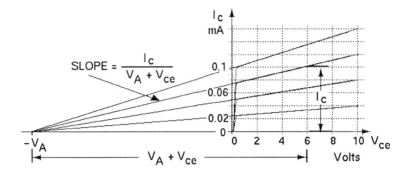

Figure 2.1   The dependence of I_C on V_{CE} is described by the Early voltage V_A.

cause collector current to be less than predicted by Equation 1.52.

As V_{BE} is increased, large numbers of electrons are injected into the base from the emitter. High level injection is defined to be when the density of electrons in the base approaches the density of acceptor atoms in the base. Extra positive voltage has to be applied to the base in order to maintain the negative charge density in the base which is produced by the high level injection of electrons from the emitter. V_{BE} is distributed between the junction and across the base region containing the high level of injected electrons. Only a portion of the voltage applied to the base and emitter terminals, V_{BE}, appears across the base emitter junction. Therefore, V_{BE} is not as effective in increasing injection across the base-emitter junction. The result is I_C proportional to $exp(V_{BE}/2V_T)$. That is, collector current does not increase as fast with increases in V_{BE} as it does in low level injection. The reduction in collector current results in a reduction in β. At low current levels, the component of base current due to spontaneous generation of electron hole pairs in the base emitter depletion region becomes significant. This component of base current varies as $e^{V_{BE}/2V_T}$. It represents base current that does not contribute to collector current. This results in a decrease in β at low current levels as shown in Figure 2.2.

2.2.3   Gummel-Poon Model

The Gummel-Poon model, like the Ebers-Moll, is not limited to positive base-emitter and positive collector-emitter voltages, but is valid for both positive and negative applied voltages. This is accomplished in a seamless way with one set of equations. Gummel-Poon was an improvement over the Ebers-Moll model in that it took into account the Early and high level injection effects.

As shown in Equation 1.53 describing an npn transistor, $I_E = -(I_{nc} +$

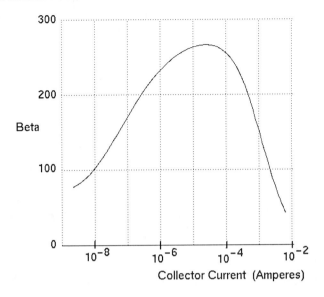

Figure 2.2   The current gain, β , of an npn transistor is shown. Gain drops off at low and high collector current. Note the logarithmic nature of the horizontal axis.

I_{pe}), and in Equation 1.54, $I_C = I_{nc} - I_{pc}$. Positive currents are defined to flow into a terminal. I_{nc} is the component of collector current due to electrons. These electrons are injected from the emitter and diffuse across the base to the collector. They contribute to both the collector and emitter currents. I_{pc} is the component of collector current due to holes injected from the base into the collector. I_{pe} is the component of emitter current due to holes injected from the base to the emitter. In this simple description where recombination in the base is considered small and ignored, I_{pe} is equal to the base current when the transistor is biased in normal forward operation with the base-emitter junction forward biased and the base-collector junction reversed biased.

From Equation 1.56

$$I_{nc} = I_s \left(e^{\frac{V_{BE}}{V_T}} - e^{\frac{V_{bc}}{V_T}} \right) \tag{2.1}$$

where

$$I_s = \frac{A_E q D_n n_i^2}{W_B N_A} \tag{2.2}$$

Define

$$I_{be1} = I_s \left(e^{\frac{V_{BE}}{V_T}} - 1 \right) \tag{2.3}$$

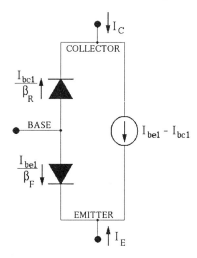

Figure 2.3 Gummel-Poon npn model without the Early effect and high level injection effects.

and

$$I_{bc1} = I_s \left(e^{\frac{V_{CE}}{V_T}} - 1 \right) \tag{2.4}$$

then

$$I_{nc} = I_{be1} - I_{bc1} \tag{2.5}$$

Also, in the normal forward operating region, I_{be1} is the collector current and I_{pe} is the base current. Therefore,

$$I_{be1} = \beta_F I_{pe} \tag{2.6}$$

The equations are symmetric so that in the reverse condition

$$I_{bc1} = \beta_R I_{pc} \tag{2.7}$$

Using Equations 2.1 thru 2.7 in Equations 1.53 and 1.54 yields the following

$$I_E = -(I_{be1} - I_{bc1}) - \frac{I_{pe}}{\beta_F} \tag{2.8}$$

$$I_C = I_{be1} - I_{bc1} - \frac{I_{ce}}{\beta_R} \tag{2.9}$$

Equations 2.8 and 2.9 are represented schematically in Figure 2.3.

Equations 2.8 and 2.9 are the Ebers-Moll model formulated in a way that allows the charge control concept used by Gummel-Poon to be included. A base charge factor, Kqb, is added to Equations 2.8 and 2.9. Kqb is a normalized number representing positive mobile charge in the

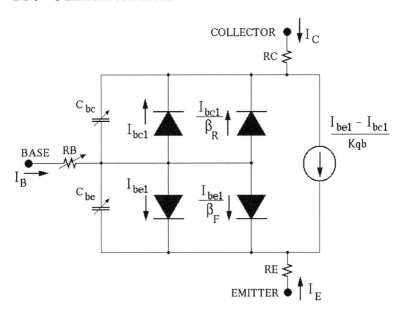

Figure 2.4 The Gummel-Poon model uses Kqb to describe high level injection and low level effects and the Early effect and the currents $Ibe2$ and $Ibc2$ to describe low level effects due to generation and recombination in the depletion regions.

base. When the collector voltage increases, Kqb becomes smaller because the base collector depletion region increases, reducing the base width and therefore the charge in the base. This is the Early effect. It causes the collector current to increase. When there is a high level of injected holes from the emitter, the extra electrons attract extra holes, increasing the positive mobile charge in the base. If the density of these added charges approaches the doping level in the base, the voltage necessary to maintain the positive charge becomes important. A portion of the applied base-emitter voltage appears across the positive charge in the base and is not available to the base-emitter junction. This is the high level injection effect. It causes collector current to be less than would be expected. High level injection is modeled as an increase in Kqb. Adding the base charge factor Kqb to Equations 2.8 and 2.9 yields

$$I_E = -\frac{I_{be1} - I_{bc1}}{Kqb} - \frac{I_{pe}}{\beta_F} \tag{2.10}$$

$$I_C = \frac{I_{be1} - I_{bc1}}{Kqb} - \frac{I_{ce}}{\beta_R} \tag{2.11}$$

Equations 2.10 and 2.11 are illustrated in Figure 2.4. Also shown in Figure 2.4 are two additional diodes carrying currents, $Ibe2$ and $Ibc2$.

These currents model base current due to recombination in the depletion regions.

Base to emitter and base to collector capacitance is also shown in Figure 2.4. These capacitances are the sum of the junction and diffusion capacitance for the junctions.

Collector and base currents are plotted in Figure 2.5, called a Gummel plot. The logarithmic vertical axis results in a straight line plot, with a slope of $1/V_T$ over a wide range. This is true since

$$I_C = I_s e^{\frac{V_{BE}}{V_T}}$$

$$ln(I_C) = ln(I_s) + \frac{V_{BE}}{V_T}$$

Since the logarithms of the collector and base currents are plotted in Figure 2.5, the log of β, the ratio of I_C to I_B, is the distance between the curves $ln(I_C) - ln(I_B)$. β decreases at both high and low values of collector current. At high levels, collector current is reduced by high level injection effects. The plot of collector current is a straight line up to about the forward knee current, IKF. At larger current values high level injection effects reduce the slope of the current plot to a value close to $V_T/2$. At low current levels, base current is larger than expected due to recombination and generation current. This current is represented by $Ibe2$ flowing in the diode in Figure 2.4. It does not contribute to the collector current. The variation of β with collector current is plotted in Figure 2.2.

2.3   MOS Transistors

MOS transistors in the 1960s could be modeled using the simple equations for hand calculations, such as Equation 1.78 discussed in Chapter 1. Model parameters corresponded to physical quantities and could be extracted from the data of simple experiments. As technology evolved, the situation became more complex due to the effects of small geometries and high fields. Model equations have become more complicated and the number of parameters required to describe effects has increased. With more complex effects to be described and larger numbers of parameters, the link between the model parameters and their physical basis has become obscure. Model parameters can be divided into two groups. *Physical* parameters that have direct physical meaning such as oxide thickness. *Electrical* parameters that are extracted from measured data but have no direct relationship to a physical quantity. Some parameters originally had physical meaning, such as junction depth, but in higher level models the parameter value is chosen to match simulator output to measured data, rather than to correspond to a physical quantity. Quoting Daniel Foty[7][page 10]

> ... models that are commonly employed can be divided into three historical generations. The first-generation models represent the oldest efforts, and come close to the ideal of describing the FET from very simple, physically based parameters. This generation consists of the Level 1, Level 2, and Level 3 models. The second-generation models introduce a very large number of empirical electrical parameters, clearly shifting the focus to the circuit design user. Extensive mathematical conditioning is introduced to improve robustness and convergence behavior of the model when used in circuit simulation, and a new approach to describing the geometry dependence, involving geometry-dependent parameters, is introduced. Due to their highly empirical nature, successful use of these models requires a tremendous amount of parameter extraction effort. This generation of models is composed of BSIM (sometimes referred to as BSIM1), HSPICE Level 28, and BISIM2. ... The development of the third generation of SPICE FET models is currently underway.

Modeling of modern MOS transistors with small feature sizes tends to be empirical rather than based on device physics. When MOS transistors were first introduced, feature sizes were large and the simple model described by Equation 1.78 was accurate. SPICE level one model for MOS transistors is accurate for large feature-size devices. There are two distinct regions of operation. The ohmic or linear region occurs for low values of drain to source voltage. The second region is called the constant current or saturation region. It occurs at higher values of drain to source voltage. Note the confusing definitions of "saturation". MOS "saturation" occurs when the voltage across the device is large, but bipolar "saturation" occurs when the voltage across the device is low.

Since transistor feature sizes have become smaller, a number of deviations from the simple model have been observed. These deviations are to be expected from what is known of device physics, but simple equations describing their performance based on physical theories do not exist. Rather an empirical approach is taken in the development of models.

PMOS transistors are complements to NMOS transistors. An NMOS transistor can be formed by diffusing N^+ source and drains into a p-well. The pwell is the body or substrate of the transistor. A PMOS transistor is the compliment of the NMOS transistor. For PMOS discussions all p and n diffusions are switched and currents and voltages are reversed. Otherwise, the descriptions are identical.

Channel Length Modulation

Small devices operating in saturation (constant current region) show an increase in drain current with drain to source voltage. This can be attributed to a decrease in the channel length, L. Increases in the drain voltage appear as increases in the reverse bias of the drain to body pn junction. This increases the width of the depletion region. Since the length of the channel is reduced by the depletion region, as the drain voltage increases, the channel length decreases. A smaller channel length results in a larger drain current. In the level 1 model, the SPICE model parameter λ is introduced to describe channel length modulation. The equation for the drain current in the linear region for $V_{GS} > V_{th}$ and $V_{DS} < V_{GS} - V_{th}$ is

$$I_{DS} = KP \frac{W}{L - 2LD} \left(V_{GS} - V_{th} - \frac{V_{DS}}{2} \right) V_{DS}(1 + \lambda V_{DS}) \qquad (2.12)$$

where L is the drawn length, LD is the lateral diffusion, $KP = \mu_n Cox$, and V_{th} is the threshold voltage. The lateral diffusion of the source and drain reduces the channel length by an amount $2LD$.

In the saturation (constant current) region $V_{GS} > V_{th}$ and $V_{DS} > V_{GS} - V_{th}$ the drain current is

$$I_{DS} = \frac{KP}{2} \frac{W}{L - 2LD} (V_{GS} - V_{th})^2 (1 + \lambda V_{DS}) \qquad (2.13)$$

λ (LAMBDA) is a SPICE parameter that approximates the increase in drain current with drain to source voltage as a linear function.

Barrier Lowering

Barrier lowering is a term used to describe the reduction of the threshold voltage as the transistor length decreases. When the transistor length becomes small, the depletion regions associated with the source and drain extend into a larger fraction of the length. This raises the surface potential making the channel more attractive for electrons, effectively reducing the threshold voltage. The barrier to electrons is lowered. Normally as the gate voltage is increased, the holes in the channel are depleted, then, with further gate voltage increases, electrons are attracted to the channel. The encroachment of the depletion regions on the channel assists in the process by increasing the channel voltage causing the depletion of holes. This is referred to as *drain induced barrier lowering* (DIBL).

Charge Sharing

Charge sharing is used to model the influence of source and drain voltages and transistor length on the threshold voltage in small MOS transistors. The threshold voltage is the gate voltage required to deplete the channel of holes and attract mobile electrons. All four regions, the gate, the substrate, the source and the drain, affect the channel surface potential and therefore the threshold voltage. The substrate is sometimes referred to as the back gate. The body effect is the decrease in threshold voltage as the body becomes more positive with respect to the source. The body effect is not limited to small devices. In addition to decreases in the threshold voltage due to increased body voltage, increases in the drain voltage also decreases the threshold voltage. In small devices, the decrease in threshold voltage with increasing drain voltage is modeled by assuming the charge in the depletion region is shared by the gate and the drain. This reduces the responsibility of the gate in maintaining the depletion region and therefore reduces the threshold voltage.

Velocity Saturation

Current flow is proportional to the drift velocity of carriers. The electric field produced by an applied voltage accelerates carriers. Accelerating carriers collide with the lattice losing their acquired momentum. This process results in an average velocity called the drift velocity. As the electric field increases, in response to an increase in applied voltage, the drift velocity increases. The drift velocity is proportional to the electric field $v_d = \mu \mathcal{E}$. The proportionality constant μ is the mobility. The result is Ohm's law where current is proportional to voltage. However, when the electric field approaches a critical field of about 1.5E4 V/cm, current no longer increases linearly with voltage. The velocity saturates at a value close to the thermal velocity for carriers in silicon. Velocity saturation occurs in small devices at low applied voltages. 1.5 V applied over a distance of 1 micron produces the critical field of 1.5E4 V/cm. Models for small devices have to include the effect of velocity saturation.

Velocity saturation can be modeled as a mobility that varies inversely with drain to source voltage.

$$\mu = \frac{\mu_0}{1 + \frac{V_{DS}}{V_{SAT}}} \tag{2.14}$$

where μ_0 is the low voltage mobility and V_{SAT} is the drain to source voltage at which the mobility has decreased by 50%.

Hot Carrier Effects

In spite of the velocity saturation mechanism that limits the drift velocity of carriers to the thermal velocity, a small fraction of carriers will acquire high energies in the large electric fields present in small devices. Since these electrons have energies greater than the average thermal energy, they are called "hot carriers". The high fields occur in the drain depletion region. Some hot carriers will collide with silicon atoms inducing "impact ionization" that produces electron-hole pairs. Electrons contribute to the drain current. The channel depletion field moves the holes into the substrate where they contribute to drain-substrate current. Some electrons may acquire enough energy to tunnel across the oxide to the gate where they contribute to gate current. Some get trapped in the oxide where they create a potential that alters the threshold voltage, resulting in device degradation with time. Hot carrier effects are reduced as the power supply voltage is reduced.

Mobility Variation

In bulk silicon, mobility is determined by thermal scattering from lattice vibrations and Coulomb scattering from ionized impurity atoms. In an MOS transistor, current flow in the channel is at the surface of the silicon at the oxide interface. Here, there is additional Coulomb scattering due to charges trapped in surface states and charges trapped in the oxide. Surface roughness also scatters carriers. As the gate voltage is increased, carrier electrons are drawn closer to the surface where surface roughness has a greater effect on mobility. Mobility decreases with increasing gate voltage. The drain voltage also affects the normal electric field pulling the electrons to the surface. Increasing the drain voltage reduces the normal field, pulls the electrons away from the surface and increases mobility.

The Variation of Threshold Voltage with Channel Width

When the channel width is less than 5 or 6 microns, it is comparable to the width of the depletion region under the gate [5]. A component of the threshold voltage is the voltage required to support the depletion region charge, Q_B. The depletion region extends out beyond the gate width. This increases Q_B and therefore the threshold voltage. The increase is more significant when the gate width is comparable to the width of the depletion region.

2.3.1   Bipolar SPICE Implementation

SPICE implements the Early effect, high level injection and low level effects using the following set of equations to describe the model shown in Figure 2.4.

$$I_C = area \left(\frac{Ibe1}{Kqb} - \frac{Ibc1}{Kqb} - \frac{Ibc1}{BR} - Ibc2 \right) \tag{2.15}$$

$$Ibe1 = IS \, exp \left(\frac{V_{BE}}{NF \, V_T} - 1 \right) \tag{2.16}$$

$$Ibe2 = ISE \, exp \left(\frac{V_{BE}}{NE \, V_T} - 1 \right) \tag{2.17}$$

$$Ibc1 = IS \, exp \left(\frac{V_{bc}}{NR \, V_T} - 1 \right) \tag{2.18}$$

$$Ibc2 = ISC \, exp \left(\frac{V_{bc}}{NC \, V_T} - 1 \right) \tag{2.19}$$

$$Kqb = Kq1 \frac{1 + (1 + 4Kq2)^{NK}}{2} \tag{2.20}$$

where $Kq1$ describes the Early effect

$$Kq1 = \frac{1}{1 - \frac{Vbc}{VAF} - \frac{Vbe}{VAR}}$$

and $Kq2$ describes high level injection

$$Kq2 = \frac{Ibe1}{IKF} + \frac{Ibc1}{IKR}$$

Theoretical expressions have been modified by the addition of a number of SPICE parameters to better fit the theory to experimental data. The Gummel plot shown in Figure 2.5 is useful in extracting SPICE model parameters from measured data. The Gummel plot is a plot of the log of the collector and base currents as a function of V_{BE}. Since these currents vary exponentially with V_{BE}, logs of the currents vary linearly with V_{BE}. The saturation current IS is the projection of the collector current on the $V_{BE} = 0$ axis. The log of the maximum ideal forward beta BF is the difference between the I_C and the I_B curves as shown in Figure 2.5. NF, the forward current emission coefficient (ideality factor), represents any departure from the ideal slope, $1/V_T$, for I_C. IKF is the high current where the slope of I_C changes. ISE, the base-emitter leakage saturation current, is the projection of the low current I_B with the $V_{BE} = 0$ axis. The slope of the base-emitter leakage

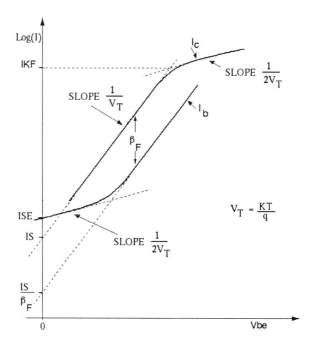

Figure 2.5 The plot of the log of the collector and base currents as a function of the base-emitter voltage is useful in determining a number of SPICE parameters.

current (low current I_B) is determined by NE. These forward parameters are important when the transistor is biased in the normal forward (active) region, with the base-emitter junction forward biased and the base-collector junction reversed biased. The transistor will operate if the functions of the collector and emitter are reversed and the base-collector junction is forward biased and the base-emitter junction is reversed biased. Usually the performance is poor but it can be characterized by the SPICE reverse model parameters. The Early voltage VAF is more easily determined from the I_C vs. V_{CE} characteristic as shown in Figure 2.1.

Bipolar Transistor DC SPICE Parameters

Parameter	Description	Units	Default
IS	Saturation current	amp	1E-16
BF	Ideal maximum forward beta		100
NF	Forward current emission coefficient		1
VAF	Forward Early voltage	volt	infinite
IKF	Corner for forward-beta high current roll-off	amp	infinite
ISE	Base-emitter leakage saturation current	amp	0
NE	Base-emitter leakage emission coefficient		1.5
BR	Ideal maximum reverse-beta		1
NR	Reverse current emission coefficient		1
VAR	Reverse Early voltage	volt	infinite
IKR	Corner for reverse-beta high current roll-off	amp	infinite
ISC	Base-collector leakage saturation current	amp	0
NC	Base-collector leakage emission coefficient		2
NK	High current roll-off coefficient		0.5
RE	Emitter ohmic resistance	ohm	0
RB	Zero-bias base resistance	ohm	0
RBM	Minimum base resistance	ohm	RB
IRB	Current at which RB falls halfway to RBM	amp	infinite
RC	Collector ohmic resistance	ohm	0

2.4   Simple Small Signal Models for Hand Calculations

Although transistors and diodes are nonlinear, linear circuit theory is useful in describing a number of circuit properties such as gain and input and output impedances. Linear analysis only works for small variations about a DC operating point. The powerful methods of linear circuit theory allow the response of a circuit to small signals to be determined. Since small signal circuit parameters depend on the DC operating point, the first step in an analysis is to determine the DC currents and voltages.

2.4.1   Bipolar Small Signal Model

A simple small signal model for a bipolar transistor can be found using the exponential dependence of collector current on base-emitter voltage in the normal operating range.

$$I_C = I_S exp \left[\frac{V_{BE}}{V_T} \right] \tag{2.21}$$

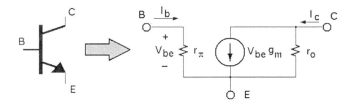

Figure 2.6   Simple small signal model for a bipolar transistor.

The transconductance shown in Figure 2.6, g_m, is defined as the change in I_C with V_{BE} when V_{CE} is constant. From Equation 2.21 it follows

$$g_m = \frac{dI_C}{dV_{BE}} = \frac{I_C}{V_T} \tag{2.22}$$

Small changes in collector current $\Delta I_C = i_c$ are approximately equal to g_m times small changes in base emitter voltage $\Delta V_{BE} = v_{be}$. With lower case used to denote small changes

$$i_c = g_m v_{be} \tag{2.23}$$

where i_c and v_{be} are the small signal collector current and base emitter voltage respectively.

Other small signal parameters shown in Figure 2.6 are r_π and r_o, the transistor input and output impedances respectively.

$$r_\pi = \frac{\partial V_{BE}}{\partial I_B} = \frac{\partial V_{BE}}{\partial I_C} \frac{\partial I_C}{\partial I_B} \tag{2.24}$$

$$r_\pi = \frac{\beta}{g_m} \tag{2.25}$$

2.4.2   Output Impedance

The output impedance r_o describes changes in I_C with V_{CE} when V_{BE} is constant. This situation can be seen in Figure 2.1 where the transistor is biased at $I_C = 0.1$ mA and $V_{CE} = 6$ V. With V_{BE} held constant, the change in I_C with V_{CE} is described by the slope of the constant V_{BE} line. These lines appear to intersect the V_{CE} axis at a single point, called the Early voltage $-V_A$. The slope of the constant V_{BE} lines is

$$g_o = \frac{I_C}{V_A + V_{CE}} \tag{2.26}$$

since V_A is usually much greater than V_{CE}

$$r_o = \frac{1}{g_o} \approx \frac{V_A}{I_C} \tag{2.27}$$

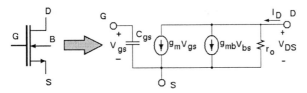

Figure 2.7   Simple small signal model for the MOS transistor.

Increasing the voltage across the transistor V_{CE} results in an increase in transistor current I_C. The physical cause is a decrease in the width of the base. As V_{CE} is increased, the reverse voltage on the collector-base pn junction increases. The collector-base depletion region extends further into the base, effectively reducing the base width. Since collector current varies inversely with base width, collector current increases.

2.4.3   Simple MOS Small Signal Model

A simple small signal model for the MOS transistor operating in saturation is shown in Figure 2.7. It ignores high order effects, but it is useful for hand calculations. The model is characterized by four parameters g_m g_{mb}, r_o, and C_{cs}.

The transconductance, g_m, can de derived from the equation for I_D in the saturation region

$$I_D = \frac{W}{L}\frac{KP}{2}(V_{gs} - V_{th})^2[1 + \lambda V_{DS}] \tag{2.28}$$

where the term involving λ has been added to the basic equation to provide a dependence of I_D on V_{DS}:

$$g_m = \frac{dI_D}{dV_{gs}} = \frac{W}{L}KP(V_{gs} - V_{th}) = \sqrt{\frac{W}{L}2I_DKP} \tag{2.29}$$

where $KP = \mu_n C_{OX}$ is the transconductance parameter, W and L are the transistor width and length, V_{th} is the threshold voltage, μ_n is the electron mobility, and C_{OX} is the gate to channel capacitance per unit area.

The output conductivity

$$g_o = 1/r_o = \frac{\partial I_D}{\partial V_{DS}}$$

$$= \frac{W}{L}\frac{KP}{2}(V_{gs} - V_{th})^2\lambda \approx \lambda I_D$$

$$r_o = \frac{1}{\lambda I_D} \tag{2.30}$$

There is an additional current source shown in the model in Figure 2.7. It accounts for the effect of the body voltage on the drain current. The body acts as a "back gate." For the NMOS transistor considered here, drain current increases when the body to source voltage increases. This is described in terms of a decrease in the threshold voltage. The dependence of threshold voltage on body to source voltage is called the "body effect". Threshold voltage is covered more fully by Tsividis[6]. Equations 1.74 and 1.75 in Section 1.6 can be used to write the threshold voltage

$$V_{th} = V_{TO} + \gamma\sqrt{2\phi_f - V_{BS}} - \sqrt{2\phi_f} \qquad (2.31)$$

V_{TO} is the zero bias threshold voltage.

The body transcondance g_{mb} describes the change in drain current with changes in body voltage

$$\frac{\partial I_D}{\partial V_{BS}} = \frac{\partial I_D}{V_{th}}\partial V_{th}\partial V_{BS}$$

$$\frac{I_D}{\partial V_{th}} = -g_m$$

$$g_{mb} = \frac{g_m\gamma/2}{\sqrt{2\phi_f V_{BS}}} \qquad (2.32)$$

If the bulk is held at a constant voltage with respect to the source, there is no change in drain current due to a change in bulk to source voltage and g_{mb} can be considered zero.

2.5   Chapter Exercises

1. In an MOS transistor circuit having a 5 V power supply, estimate the channel length L that will give rise to velocity saturation. Assume a threshold voltage of 1 V, a surface mobility of 500 $cm^2/V.s$ and a saturation velocity of $1.5x10^7 cm/sec$.

2. Threshold voltage increases when substrate doping increases. T F

3. Threshold voltage increases when the substrate bias voltage increases. T   F

References

[1] P.R. Gray and R.G. Meyer, *Analysis and Design of Analog Integrated Circuits*, 2nd edition, Wiley, New York, c. 1984, pp. 1-5.

[2] R.S. Muller and T.I. Kamins, *Device Electronics for Integrated Circuits*, 2nd edition, Wiley, New York, c. 1986, pp. 15-27, 173-188, 235-244.

[3] K. Lee, M. Shur, T.A. Fjeldy and T. Ytterdal, *Semiconductor Device Modeling for VLSI*, Prentice Hall, Englewood Cliffs, NJ, c. 1993, p. 63.

[4] MicroSim Corporation, *Circuit Analysis Reference Manual*, Version 6.0, MicroSim Corp., Irvine, CA., 1994.

[5] Giuseppe Massobrio and Paolo Antognetti, *Semiconductor Device Modeling with SPICE*, 2nd edition, McGraw-Hill, New York, 1993.

[6] Yannis Tsividis, *Operation and Modeling of The MOS Transistor*, 2nd edition, McGraw-Hill, New York, 1999.

[7] Daniel Foty, *MOSFET MODELING WITH SPICE Principles and Practice*, Prentice Hall PTR, Upper Saddle River, NJ 07458, 1997.

chapter 3

Current Sources

Current sources are the foundation of circuit design in microelectronics. Current sources provide biasing for circuit operation. They serve as output drivers. They serve as load elements in amplifier input stages. Even logic gates can be modeled as a collection of variable current sources. Analysis of circuits in proceeding chapters will often show a resistance biasing the block under analysis. Current mirrors are used almost exclusively for this purpose in microelectronics. Current sources offer the advantages of smaller size, higher accuracy and can be designed to provide temperature coefficients of current as needed. However, resistors can and do serve well as current sources in some instances. Let us first consider the characteristics of an ideal DC current source as provided in circuit simulators such as SPICE.

- Constant current of any value is provided at all times.

- Infinite output impedance means there is no change in the source current value due to changes in the output node voltage.

- The source has infinite compliance, and will provide the specified current regardless of the voltage across the source.

- An ideal current source can either sink or source current. The polarity of the specified DC current and the nodal connection of the current source to the rest of the circuit determine how the source behaves. (Most simulators have an ideal current source with two nodes: positive and negative. Positive current flow in the ideal source is defined as flowing into the positive node and out of the negative node.)

Unfortunately, physical constraints apply in the real world of semiconductors, and real current sources fall short of perfection.

- Current provided by integrated circuit current sources are constant within some tolerance, and the value of current depends on limitations of device size, power dissipation and process Early voltages.

- Source output impedance values typically fall between several hundreds of kilohms and several megohms. The voltage appearing across the source modifies current.

- Real supplies power circuits, and every circuit element has a voltage drop across it due to the internal resistance. The minimum compliance voltage a current source can realize is limited by its internal resistance, and the maximum compliance voltage is limited by the circuit's supply and process breakdown voltages.

- We can build current sources and current sinks, but the same element typically doesn't perform both functions under normal operating conditions.

It is a fact of life that we have physical constraints that forbid the existence of ideal elements. Fortunately, the design of integrated circuit current references that are "good enough" occurs every day.

Let us start our discussion of current sources with the simplest example: the resistor. If we connect a resistor R between a voltage supply V and a load, the current supplied to the load is dependent on the value of V, the value of R and the impedance of the load:

$$I = \frac{V}{R + R_{load}}$$

In reality, resistor R will have some temperature coefficient and possibly a voltage coefficient. The absolute tolerance of the resistor will have a manufacturing tolerance associated with it. A resistor may seem to be a poor choice for a current source, but it may be that a resistor is perfectly adequate. For example, if the requirement is for a minimum current to be provided to the load, the tolerances on V and on R_{load} can be considered, and R sized appropriately to always provide the minimum current.

As previously stated, current sources used in microelectronics are usually made with transistors in a configuration known as the current mirror. In a current mirror, one transistor serves as a reference device. Current is supplied to the reference device and the reference device generates a value of V_{be} in bipolar technology, or a value of V_{gs} in MOS technology. These reference voltages are then provided to other transistors that serve to recreate, or mirror, the initial current. Let us begin our discussion of transistor-based current sources with the bipolar case, and then recreate the discussions for MOS.

3.1   Current Mirrors in Bipolar Technology

The relationship between collector current and base-emitter voltage in bipolar transistors operating in their linear region is defined by what is commonly known as the diode equation:

$$I_c = I_s e^{\frac{V_{be}}{V_T}} \tag{3.1}$$

where I_s is a process parameter called saturation current and V_T is thermal voltage. If a particular value of V_{be} is placed across the BJT's base-emitter junction, a particular value of collector current will flow. Conversely, if we somehow force a collector current to flow, a particular value of V_{be} will be created across the base-emitter junction. We can use the diode equation to create current mirrors. Refer to Figure 3.1a.

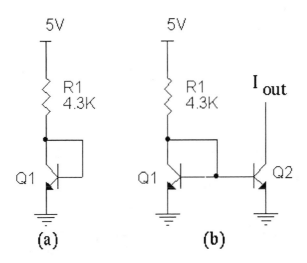

Figure 3.1   Simple Current Mirror.

Here, NPN transistor Q1 is diode-connected. If we assume Q1 is an ideal transistor with $\beta = \infty$, then there is no base current, and all the current flows in the collector of Q1. The value of this current can be initially estimated using the approximation of $V_{be} = 0.7V$

$$I_c(Q1) = \frac{5 - 0.7}{4.3k} = 1mA \tag{3.2}$$

We can then calculate the value of V_{be} present by using the diode equation with a known value of I_s. The calculated value of V_{be} may be returned into Equation 3.2 and another solution for $I_c(Q1)$ obtained. Several iterations of this procedure will result in solutions for both collector current and base-emitter voltage that are fairly accurate. Refer now to Figure 3.1b. Here, we have placed a second transistor such that $V_{be}(Q2) = V_{be}(Q1)$. If we assume that $Q1$ and $Q2$ are identical in all respects, then $I_s(Q1) = I_s(Q2)$, and ultimately $I_c(Q2) = I_c(Q1)$. This is the basic principle of operation for a current mirror.

Example

Using the circuit from Figure 3.1b, find the collector current in transistor $Q2$ if $R_1 = 10$ $k\Omega$. Use VCC $= 5$ V, $I_s = 200E - 18$ A and $V_T = 26$ mV. Assume ideal NPN transistors with $\beta = \infty$.
Using the approximation $V_{be} = 0.7V$, solve for $I_c(Q1)$:

$$I_c(Q1) = \frac{5 - 0.7}{10,000} = 430 \ \mu A$$

Find the "real" V_{be} value:

$$V_{be} = 26mVln \left[\frac{430E - 6A}{200E - 18A} \right] = 738.3 \ mV$$

Recalculate the current:

$$I_c(Q1) = \frac{5 - 0.7383}{10,000} = 426.2 \ \mu A$$

Recalculate Vbe:

$$V_{be} = 26mVln \left[\frac{426.2E - 6A}{200E - 18A} \right] = 738.1 \ mV$$

Another iteration may be made, but the change in current between iterations was only 1%. This level of refinement is usually good enough for first-pass design. Based on our assumption that both transistors are ideal, we can conclude that the collector current in $Q2$ is equal to that in $Q1$ and so $I_c(Q2) = I_c(Q1) = 426.2 \ \mu A$.

We can expand this analysis to multiple transistors. Consider the circuit in Figure 3.2a. This circuit has two mirror transistors. Using the same assumptions of ideality and identical transistors, we come to the conclusion that each mirror transistor is sinking current equal to the reference current. Furthermore, if we tie the collectors of both mirror transistors together, the output current of the mirror is equal to twice the reference current, as shown in Figure 3.2b.

This leads to an interesting point. What happens if transistors $Q2$ and $Q3$ are merged into a single device? Making the emitter size twice as big as in Q1 can do this. The correct answer is that the output current of the "2X" mirror will be twice the reference current. Accuracy of the

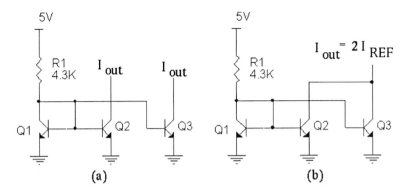

Figure 3.2 Multiple transistor current mirror.

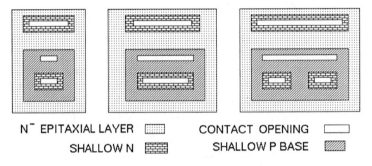

N⁻ EPITAXIAL LAYER   CONTACT OPENING

SHALLOW N   SHALLOW P BASE

Figure 3.3 NPN current mirror layout. Blue indicates shallow-n+ doping for emitter and collector ohmic contact. Green indicates shallow-p base. White indicates contact openings. Light yellow indicates n- epitaxial layer.

mirror will depend on the physical layout of the transistors. Figure 3.3 shows two options for "2X" layout.

For current multiplication by an integer value, mirror 2 will be more accurate. This is because layout and fabrication gradients should affect the base-emitter junctions of the reference and mirror 2 in a similar manner. Effects on mirror 1 will be somewhat different. However, for current multiplication by a fractional value, say 1.5X, mirror 1 can be laid out to provide the additional current by increasing the emitter area to be a 1.5X multiple of the reference's emitter area. Once again, we have assumed ideal transistors. Let us consider the effect of finite forward current gain β on the accuracy of our current mirror.

β is defined as I_c/I_b. A typical range of values is $100 < \beta < 400$. Thus, for any current to flow in the collector, some current must be flowing in the base. If our circuit is based on a diode-connected transistor, the base current will be subtracted from the collector current and an error will result. Figure 3.4a shows the current mirror provided with an ideal

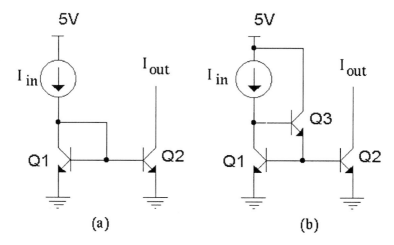

Figure 3.4   Multiple transistor current mirror.

reference current I_{in}. Since we now have a finite forward current gain, the currents flowing in the bases of $Q1$ and $Q2$ are also supplied by I_{in}. These currents reduce the amount of I_{in} that flows in the collector of $Q1$. If we approximate the base currents of both $Q1$ and $Q2$ as equal, we have

$$I_{out} = I_c(Q2) = I_{in} - 2\frac{I_c(Q2)}{\beta} \tag{3.3}$$

$Q1$'s base-emitter voltage will reflect the amount of collector current flowing, and $Q2$ will mirror a current that is less than I_{in}.

Figure 3.4b shows a circuit that reduces the error due to base current. Transistor $Q3$ acts as a buffer and provides the base current for $Q1$ and $Q2$. The emitter current for $Q3$ is then equal to

$$I_E(Q3) = I_b(Q1) + I_b(Q2) = \frac{2I_c(Q1)}{\beta} \tag{3.4}$$

The base current of $Q3$ is then given as

$$I_b(Q3) = \frac{I_E(Q3)}{\beta + 1} = \frac{2I_c(Q1)}{1 + \beta(\beta + 1)} \tag{3.5}$$

If we approximate $I_{in} = I_c(Q1)$, we can then say

$$\frac{I_{out}}{I_{in}} = 1 - \frac{2}{\beta^2 + \beta} \tag{3.6}$$

The error increases with the number of mirror transistors connected to the base rail and decreases with increasing current gain.

Example

Use the circuits in Figure 3.4 to determine the value of I_{out} if $I_{in} = 50\mu A$ and $\beta = 100$. Assume all transistors are identical. Let's start with Figure 3.4a. Since the two transistors are identical, we can assume that whatever collector current exists in $Q1$ will be mirrored in $Q2$. Thus, $I_c(Q1) = I_c(Q2) = I_c$. Next, we can assume that β is identical, so base currents will be identical: $I_b(Q1) = I_b(Q2) = I_b$. Now we can use Kirchoff's current law at the collector of $Q1$ to obtain

$$I_c = I_{in} - \frac{2I_c}{\beta} \tag{3.7}$$

This can be rewritten as

$$I_c = \frac{I_{in}}{1 + \frac{2}{\beta}} \tag{3.8}$$

Thus, $I_{out} = I_c = 50\mu A/1.02 = 49.61\mu A$.

Now, with the circuit in Figure 3.4b, we have the following relationships:

$$I_c(Q1) = I_c(Q2) = I_c = I_{out}$$

and

$$I_b(Q1) = I_b(Q2) = I_b$$

For $Q3$, we have $I_E(Q3) = 2I_b$ and $I_b(Q3) = 2I_b/(\beta + 1)$. Again using Kirchoff's current law at the collector of $Q1$, we obtain

$$I_{in} = I_c + I_b(Q3) = I_c + \frac{2I_b}{\beta} = I_c + \frac{2I_c}{\beta^2 + \beta}$$

Rewritten, we have

$$I_c = \frac{I_{in}}{1 + \frac{2}{\beta^2 + \beta}} \tag{3.9}$$

Thus, $I_{out} = 49.990\mu A$.

We have seen how current gain can be accomplished by using multiples of emitter area in the mirror transistor. This is a simple extension of the diode equation. A change in V_{be} of 18 mV results in a doubling of collector current (proof is left as an exercise for the student). Similarly, changing the emitter area of a transistor can be viewed as directly scaling the I_s parameter. If A_E is scaled by a factor of 2, then I_s for that transistor scales by a factor of 2. The same effect can be created using a resistor.

Consider the circuit in Figure 3.5. Given particular values of I_{in} and R, the voltage drop developed across R will increase the V_{be} of $Q2$ with

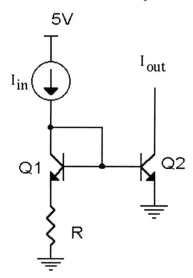

Figure 3.5   Current mirror with output current gain.

the result that I_{out} will be greater than I_{in}. Every multiple of 18 mV will result in I_{out} being a factor of 2 greater than I_{in}. For example, if $R = 360\Omega$ and $I_{in} = 50\mu A$, the voltage drop across R will be 18 mV, and I_{out} will be approximately twice I_{in}.

Example

For the circuit in Figure 3.5, assume $\beta = 100$, $I_s = 200E - 18A$, $I_{in} = 100\mu A$ and the desired value of I_{out} is $150\mu A$. Find the required value of R.

We know the collector current of $Q2$ will be $150\mu A$. Base current in $Q2$ will then be $1.5\mu A$. The collector current in $Q1$ is then given by

$$I_c(Q1) = 98.5\mu A - I_b(Q1) = 98.5\mu A - \frac{I_c(Q1)}{\beta}$$

or

$$1.01 I_c = 98.5\mu A$$

This gives $I_c(Q1) = 97.525\mu A$. Using Kirchoff's Voltage Law at the bases of $Q1$ and $Q2$, we find

$$V_{be}(Q2) = V_{be}(Q1) + 97.525\mu A\ R$$

Now

$$V_{be}(Q1) = V_T ln\left[\frac{97.5E - 6}{200E - 18}\right] = 0.6997V$$

and

$$V_{be}(Q2) = V_T ln \left[\frac{150E - 6}{200E - 18} \right] = 0.7109V$$

Solving the KVL equation, we find

$$R = \frac{V_{be}(Q2) - V_{be}(Q1)}{97.525\mu A} = 115\Omega$$

Note that placing a resistor in the emitter of Q2 as shown in Figure 3.6 would serve to decrease the V_{be} and would reduce the value of I_{out}. The circuits in Figures 3.5 and 3.6 are examples of Widlar current sources. They are named after Robert Widlar, one of the pioneers in transistor electronics. Solving for the required resistance from two known currents is fairly straightforward. It is slightly more difficult to find the output current from a known input with a fixed value of R.

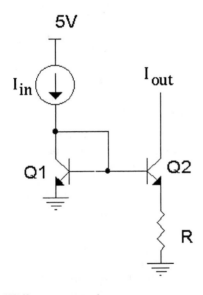

Figure 3.6   Widlar current mirror.

Example

Use the circuit in Figure 3.6 to find the value of I_{out}, if $I_{in} = 100\mu A$, $\beta = 100$, $I_s = 200E - 18A$ and $R_2 = 100\Omega$.

Let us start by approximating the base currents. We know the voltage drop across R will reduce the collector current of Q2. If Q2 carried $100\mu A$, the drop across R would be 10 mV. A change of 18 mV is required to halve the current, so we can expect current greater than $50\mu A$ to be flowing. Let us approximate $I_b(Q2)$ as $1\mu A$. Then $I_c(Q1) \approx 98\mu A$. Now we can use KVL at the transistor bases:

$$V_{be}(Q1) = V_{be}(Q2) + I_{out}R$$

Substituting the diode equation for V_{be}, we have

$$V_T ln\left[\frac{I_c(Q1)}{I_s}\right] = V_T ln\left[\frac{I_{out}}{I_s}\right] + I_{out}R$$

Rearranged, we have

$$I_{out} = \frac{V_T}{R}ln\left[\frac{I_C(Q1)}{I_{out}}\right]$$

This is a transcendental equation. I_{out} is both the solution and a variable within the problem. This requires an iterative solution. Take a first guess and solve to find a point at which the equation is an identity. The chart of values below shows the method.

"Guess-timate"	Solved value	Identity?
$50\mu A$	$18\mu A$	way off
$75\mu A$	$74.8\mu A$	not quite
$76\mu A$	$71.3\mu A$	too far
$74.9\mu A$	$75.1\mu A$	not enough
$74.95\mu A$	$74.97\mu A$	close enough

Fortunately, circuit simulators can perform these operations very quickly. However, it is good engineering practice to complete a "paper design" before simulation so that unexpected results can be checked early in the design phase.

One of the most important qualities of the ideal current source is its infinite output impedance. The ideal source provides a constant output current regardless of the voltage of the output node. Practical sources, however, have finite output resistance that must be considered.

Let us start with the mirrors in Figure 3.4. In either circuit, the output stage is a single transistor. The output resistance of the mirror is equal to the output resistance of the transistor. We know this quantity as

$$r_o = \frac{V_A}{I_c} \tag{3.10}$$

where V_A is the Early voltage. The slope with which collector current increases with increasing collector-emitter voltage is defined as the inverse of r_o. This change in current can be modeled as an extension of the diode equation:

$$I_c = I_s e^{\frac{V_{be}}{V_T}}\left[1 + \frac{V_{ce}}{V_A}\right] \tag{3.11}$$

Example

For the circuit in Figure 3.4a, we have already determined the collector current to be 49.61 μA. At what value of V_{ce} will this be true? What is I_{out} if $V_A = 100V$ and $V_{out} = 20V$?

Since the reference transistor $Q1$ has a $V_{ce} \approx 0.7V$, $V_{ce}(Q2)$ should be 0.7V for the mirror to work correctly. For $V_{ce} = V_{out} = 20V$

$$I_c(Q2) = I_s e^{\frac{V_{be}(Q2)}{V_T}} \left[1 + \frac{20}{100} \right]$$

$$I_c(Q1) = I_s e^{\frac{V_{be}(Q1)}{V_T}} \left[1 + \frac{0.7}{100} \right]$$

So $\frac{I_c(Q2)}{I_c(Q1)} = \frac{1.2}{1.07} = 1.1215$. I_{out} has increased by more than 12%. Lowering the transistor collector current can increase output resistance, but this is often not an option in a design. Early voltage is usually fairly well fixed as a result of the fabrication process. Fortunately, there are several circuit design options available that allow us to increase r_o from several hundreds of kilohms to several megohms. Consider the Widlar current mirror. Adding the resistor as shown in Figure 3.6 helps to increase output resistance. We can understand this more easily by drawing the small-signal equivalent circuit as shown in Figure 3.7.

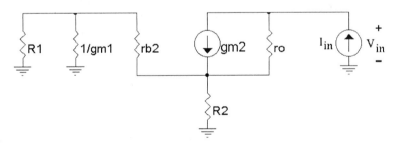

Figure 3.7   Widlar current mirror small-signal equivalent circuit.

Since $Q1$ is diode-connected it is modeled as $1/gm_1$. The quantity r_b is defined as β/gm. Since r_b is greater than $1/gm_1$ by a factor of β, the parallel combination of R_1 and $1/gm_1$ can be ignored, and the circuit reduces to that shown in Figure 3.8.

Applying test source I_{in}, we see that all the test current flows through the parallel combination of R_2 and r_{b2}. The resulting voltage at V_e is

$$v_e = -i_{in}(r_{b2} \| R_2) \tag{3.12}$$

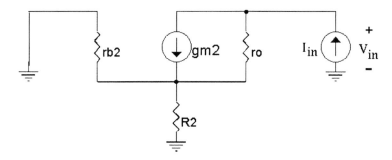

Figure 3.8 Simplified small-signal equivalent circuit for the Widlar current mirror.

Current through r_o is

$$i(r_o) = i_{in} - gm_2 v_e = i_{in} + i_{in}gm_2(r_{b2}\|R_2) \tag{3.13}$$

Voltage v_{in} is then given by the sum of the voltage drops across the two resistances:

$$v_{in} = -v_e + i(r_o)r_o \tag{3.14}$$

Output resistance is then given as

$$R_o = \frac{v_{in}}{i_{in}} = r_{b2}\|R_2 + r_o\left[1 + gm_2 r_{b2}\|R_2)\right] \tag{3.15}$$

Expanding the parallel resistance and reducing the result leads to

$$R_o = r_o \frac{1 + gm_2 R_2}{1 + \frac{gm_2}{\beta}}$$

Finally, since $gm_2 R_2 \ll \beta$, and $gm_2 = I_c(Q2)/V_T$, we obtain

$$R_o = r_o \left[1 + \frac{I_c(Q2)R_2}{V_T}\right] \tag{3.16}$$

This is a very important result because it shows that every increase of 26 mV across R_2 increases the mirror output resistance by r_o. That is, 26 mV across R_2 gives $R_o = 2r_o$, 52 mV gives $R_o = 3r_o$, etc. This result can also be extrapolated back to the simple current source. Using emitter degeneration resistors for both the reference and the mirror transistors will increase the output resistance, but without introducing current scaling effects. In general, this technique is limited by the amount of voltage dropped across resistor R_2. It is usually not desirable to have more than about 150 mV across the degeneration resistors.

Another technique to increase the output resistance is called cascoding. A cascode current mirror uses two mirrors stacked one on top of

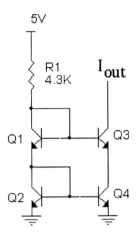

Figure 3.9   Bipolar cascoded current mirror.

the other and uses the high output resistance of the bottom mirror to increase the output resistance of the top mirror as shown in Figure 3.9.

If we assume that the base voltages do not change with variation of $Q3$'s collector voltage, we can use 3.18 with $r_o(Q3)$ substituted for R_2:

$$Ro \approx r_o \left[\frac{1 + gm_2 r_o(Q3)}{1 + \frac{gm_2 r_o(Q3)}{\beta}} \right] \qquad (3.17)$$

Thus, R_o can be increased by a factor of β by cascoding.

It is important to note here that our assumption in this analysis is flawed. As the collector voltage of $Q3$ varies, Early voltage effects cause changes in the collector current. This requires $V_{be}(Q3)$ to change slightly to maintain constant current. A thorough small-signal analysis of the cascode current source shows an output resistance increase of only $\beta/2$.

The Wilson current mirror shown in Figure 3.10 is a variation on the cascode theme. This circuit uses a negative feedback approach to maintain a well-regulated output current. Base current cancellation is also provided, making this circuit relatively insensitive to changes in β.

Base current in $Q2$ is multiplied by $\beta + 1$ and exits $Q2$'s emitter. Current flowing in the collector of $Q3$ causes $Q1$ to mirror the same current. If $Q2$ begins to provide too much current, the mirror action of $Q1$ and $Q3$ decreases the available drive to $Q2's$ base and limits the current. If β is constant across all three transistors, base current cancellation is achieved and a well-regulated output current is provided.

The voltage drop across $Q1$ is equal to $V_{be}(Q2) + V_{be}(Q3)$, while $Q3$ is limited to $V_{ce} = V_{be}$. Thus, Early voltage effects can be ignored. Modulation of $Q2's$ collector voltage will have very little effect on the value of output current, which implies a high output resistance. Small

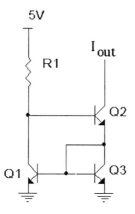

Figure 3.10   Wilson current mirror.

signal analysis yields

$$R_o = \frac{\beta r_o}{2} \tag{3.18}$$

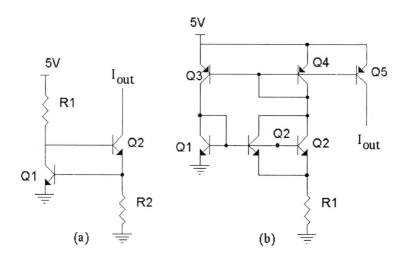

Figure 3.11   A. Vbe/R current reference.   B. (delta)Vbe/R current reference.

Two common current references are shown in Figure 3.11. The V_{be}/R current source is shown in Figure 3.11a, and the $\Delta V_{be}/R$ source is shown in Figure 3.11b.

The V_{be}/R current source takes its name from its transfer function:

$$I_{out} = \frac{V_{be}(Q1)}{R_2} = \frac{V_T}{R_2} ln \left[\frac{I_c}{I_s} \right] \qquad (3.19)$$

The output current is largely independent of supply voltage as long as sufficient current is available to turn $Q1$ on. However, large changes in $Q1's$ collector current will change its V_{be} and can influence the output current value. The temperature coefficient associated with I_{out} will be negative since V_{be} decreases with temperature while integrated resistors typically increase in value.

The $\Delta V_{be}/R$ source makes use of the thermal voltage to establish the output current. The current in $Q1$ is mirrored to $Q2$. $Q2$ has an emitter area twice that of $Q1$ with the result that the current density in $Q2$ is half that of $Q1$. This results in $V_{be}(Q2)$ being lower than $V_{be}(Q1)$. The difference is called ΔV_{be} and is dropped across resistor R_1 to set the output current:

$$I_{out} = \frac{V_T ln \left[\frac{A_2}{A_1} \right]}{R} \qquad (3.20)$$

In this case, $Q1$, $Q2$ and R_1 set up the output currents, but the mirror output is taken from the PNP transistor current rail.

3.2   Current Mirrors in MOS Technology

MOS devices can be used to build current mirrors in direct analogue to the bipolar cases. MOS devices operate linearly in the saturation region. This requires $V_{gs} > V_{th}$ and $V_{ds} \geq V_{gs} - V_{th}$. The equation defining MOS device operation in the saturation region is

$$I_d = \frac{W}{L} \frac{\mu C_{ox}}{2} (V_{gs} - V_{th})^2)[1 + \lambda(V_{ds} - (V_{gs} - V_{th}))] \qquad (3.21)$$

Setting $\mu C_{ox} = KP$ and the last instance of $V_{gs} - V_{th} = V_{ds_{sat}}$, Equation 3.21 reduces to

$$I_d = \frac{W}{L} \frac{KP}{2} (V_{gs} - V_{th})^2 [1 + \lambda(V_{ds} - V_{ds_{sat}})] \qquad (3.22)$$

Current flows in $M1$ as a result of V_{gs1}. If $M2$ is in saturation, and if $W_2/L_2 = W_1/L_1$, then I_{out} will be equal to I_{d1}. In MOS technology, scaling of currents is easily accomplished by manipulating the W/L ratios of each transistor. However, it is important to keep all the mirror devices operating in the saturation region to maintain proper operation. The minimum voltage across the mirror transistor is $V_{ds_{sat}} = V_{gs} - V_{th}$.

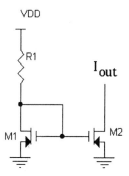

Figure 3.12   MOS current mirror.

Under these conditions, the output resistance of the current mirror is the output resistance of the mirror transistor:

$$R_o = \frac{1}{\lambda I_d} \tag{3.23}$$

where λ is the channel length modulation parameter.

Five variables are available as design parameters: W_1, L_1, W_2, L_2 and V_{gs}. Normally, values of L and V_{gs} are picked first to simplify the design process. For example, making all values of L equal reduces the current ratio equation to a ratio of transistor widths:

$$\frac{I_{d2}}{I_{d1}} = \frac{W_2}{W_1} \tag{3.24}$$

Also, making all values of L the same serves to make the effects of process variations constant from transistor to transistor. Lateral diffusion, etch effects and photolithography errors will then affect the circuit in a "common mode" manner. Errors tend to cancel under these conditions. In general, it is a good practice to make L as large as possible for analog designs. Increasing L reduces the value of λ. Setting L equal to three times the process minimum length is a good rule to start with. This rule can be modified after you have experience with a particular process. It is also a good practice to design for a specific V_{gs} that is somewhat larger than V_{th}. Higher values of V_{gs} allow smaller values of W to be used, but the value of $V_{ds_{sat}}$ will be increased. Values of V_{gs} that approach V_{th} result in physically large transistors.

Example

Design a current mirror using n-channel MOS devices. Use the circuit shown in Figure 3.12. Assume supplies are $V_{dd} = 5V$ and ground. Both reference current and output current are to be $20\mu A$. $KP = 50\mu A/V^2$,

$L = 5\mu m$, $V_{th} = 0.8V$ and $\lambda = 0.04V^{-1}$. Use $V_{gs} = 1.3V$. Determine the minimum voltage required to stay in saturation and find the output resistance.

First, we start by calculating the value of R required to provide the reference current:

$$R = \frac{V_{dd} - V_{gs}}{20\mu A} = \frac{5V - 1.3V}{20E - 6A} = 185k\Omega$$

Since $I_{out} = I_{in}$, we can solve for W_1 and W_2 at the same time:

$$I_{d1} = I_{d2} = 20\mu A = \frac{W}{5E - 6}\frac{50E - 6}{2}(1.3 - 0.8)^2$$

From this we obtain $W = 16\mu m$. The minimum voltage required to stay in saturation is

$$V_{ds_{sat}} = 1.3 - 8 = 0.5V$$

Output resistance is found to be

$$r_o = \frac{1}{20E - 6 * 0.04} = 1.25M\Omega$$

It is important to realize that r_o of the simple current mirror is proportional to $1/I_d$, and so high output resistance is obtained only for low values of current.

MOS devices do not use a bias current, and so buffered current mirrors aren't needed in MOS technology. However, cascoded sources such as the one shown in Figure 3.13 are frequently used.

An added error term is found in MOS cascode structures. The body effect results from a non-zero potential between the source and the bulk semiconductor in which the transistor is built. The body effect results in an apparent increase in V_{th}. For our analysis, we will model the body effect as a conductance gm_b that exists in parallel with gm.

Let us also consider V_{gs} for a moment. We know that the actual value of V_{gs} must exceed V_{th} by some value in order for the transistor to be in saturation. If we define $V_{gs} = V_{th} + \Delta V$, we can solve the saturation drain current equation for ΔV:

$$\Delta V = \sqrt{\frac{2I_d L}{KPW}} \tag{3.25}$$

Note that $V_{ds} \geq \Delta V$ to keep the transistor in saturation.

Output current I_{out} is determined by the ratio of $\frac{W_4 L_2}{W_2 L_4}$ and the value of I_{ref}. If we assume all transistors in Figure 3.13 are identical, then

$$V_{gs}(M2) = V_{gs}(M4) = V_{th} + \Delta V \tag{3.26}$$

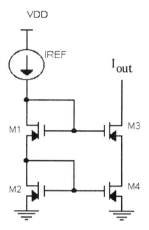

Figure 3.13   MOS cascoded current mirror.

The voltage at the gates of $M1$ and $M3$ is equal to $2\Delta V_{th} + V_{th}$. For $M3$ to remain in saturation

$$V_{gs}(M3) \geq 2(V_{th} + \Delta V) - \Delta V - V_{th} = V_{th} - \Delta V \qquad (3.27)$$

The total voltage across the cascoded mirror is then $V_{th} + 2\Delta V$. The output voltage swing for which the cascoded mirror will remain in saturation has been increased by $V_{th} - \Delta V$ over the simple mirror.

Small signal output resistance is obtained from the same procedure used in analysis of the bipolar cascoded mirror. A current I_o is forced into the cascode, the voltage V_o across the source is measured and r_0 is calculated. The AC equivalent schematic is shown in Figure 3.14a with the equivalent circuit provided in Figure 3.14b.

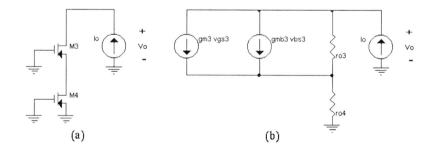

Figure 3.14   A. MOS cascoded current mirror ac-equivalent schematic.   B. Small-signal equivalent circuit for ac-equivalent schematic.

Small signal analysis gives the following equations:

$$v_4 = i_o r_{o4} \tag{3.28}$$

$$v_{gs3} = v_{bs3} = -v_4 \tag{3.29}$$

$$v_o = (i_o - gm_3 v_{gs3} - gm_b v_{bs3})r_{o3} + v4 \tag{3.30}$$

Substituting and rewriting gives

$$v_o = i_o(1 + gm_3 r_{o4} + gm_{b3} r_{o4})r_{o3} + i_o r_{o4} \tag{3.31}$$

Then

$$r_o = \frac{V_o}{I_o} = r_{o3} + r_{o4} + r_{o4} r_{o3}(gm_3 + gm_{b3}) \tag{3.32}$$

Since r_{o3} is much less than $r_{o3} r_{o4}(gm_3 + gm_{b3})$, this simplifies to

$$r_o \approx r_{o4}(1 + r_{o3}\left[gm_3 + gm_{b3}\right] \tag{3.33}$$

The total output resistance is equal to the output resistance of $M4$ multiplied by one plus the voltage gain of transistor $M3$. This result shows that output resistance can be optimized without requiring any change to the output current value. Current is set by $M2$ and $M4$, while output resistance can be increased by dealing with $M3$.

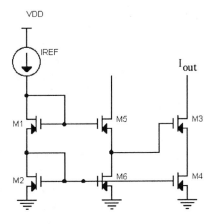

Figure 3.15   Improved MOS cascoded current mirror.

One of the biggest drawbacks in using the cascode mirror shown in Figures 3.13 and 3.14 is the increase of $V_{th} + \Delta V$ needed to keep $M3$ and $M4$ in saturation. An improved cascode current mirror can be built by inserting a voltage level shifting circuit between the reference and the output. This circuit is shown in Figure 3.15.

The key to making the circuit in Figure 3.15 work correctly is ensuring $V_{ds}(M1) = V_{th} + 2\Delta V$. If $V_{gs}(M2) = V_{th} + \Delta V$, and $V_{gs}(M1) = V_{th} + 2\Delta V$, then the voltage at the gate of $M5$ is equal to $2Vth + 3\Delta V$. The voltage at the source of $M5$ is then V_{gs} lower, or $V_{th} + 2\Delta V$.

The voltage at the drain of $M4$ is then only ΔV, and $V_{gs}(M3)$ is $V_{th} + \Delta V$. Thus, $V_{ds}(M3)$ must be greater than or equal to ΔV in order to maintain saturation and the minimum output voltage is $2\Delta V$. We have reduced the minimum saturation voltage by V_{th}.

Now, how to size the transistors? Analysis of the circuit gives the following equations:

$$I_{ref} = I_d(M1) = I_d(M2) \tag{3.34}$$

$$I_d(M1) = \frac{W_1}{L_1}\frac{KP}{2}(V_{gs1} - V_{th})^2 \quad where\, V_{gs1} = V_{th} + 2\Delta V \tag{3.35}$$

$$I_d(M2) = \frac{W_2}{L_2}\frac{KP}{2}(V_{gs2} - V_{th})^2 \quad where\, V_{gs2} = V_{th} + \Delta V \tag{3.36}$$

Substituting Equation 3.34 and Equation 3.35 into Equation 3.36 leads to

$$\frac{W_1}{L_1}(2\Delta V)^2 = \frac{W_2}{L_2}\Delta V^2 \tag{3.37}$$

This can be written as

$$\frac{W_1}{L_1} = \frac{1}{4}\frac{W_2}{L_2} \tag{3.38}$$

This provides the extra ΔV needed in $V_{gs}(M1)$. Setting $W_2/L_2 = W_4/L_4 = W_6/L_6$ sets the output current. W_5/L_5 is set to ensure $M5$ also operates in saturation while W_3/L_3 is set to optimize output resistance.

3.3   Chapter Exercises

1. Using the simple current mirror in Figure 3.1, design a circuit that has a reference current of 150 μA. Transistor saturation current, $I_s = 2E - 16$. Assume $\beta = 100$, $\beta = 400$, $\beta = \infty$. Find V_{be} to 1% accuracy. What is the percentage error in output current for the three cases above?

2. If the circuit described in problem 1 with $\beta = 250$, what is the output current variation if there were ±5% variation in the value of the supply voltage?

3. For the circuit described in problem 1 with $\beta = 250$, what is the output current variation if the manufacturing tolerance on the value of R is ±30%?

4. For the circuit in problem 1 with $\beta = 250$, what is the variation in output current if resistor R has a temperature coefficient (abbreviated as TC) of + 2500 parts per million (abbreviated as PPM) per degree Centigrade over the temperature range from $0 \deg C$ to $70 \deg C$, where $25 \deg C$ is typical? You should include the temperature variation of $-2mV/\deg C$ for V_{be} as well.

5. Problems 1 through 4 have completed a sensitivity analysis for the base design. Comment on the expected total error and the maximum possible error for this design. What should the specification be for this design over supply voltage, temperature and manufacturing tolerance? What is the major cause of the error? How can this problem be designed out?

6. Prove that a change in V_{be} of 18 mV results in a doubling in the value of I_c.

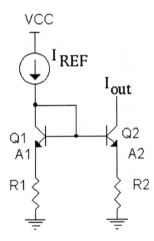

Figure 3.16   Schematic for exercise 7.

7. Use the schematic shown in Figure 3.16. Expand the diode equation to obtain an equation that accounts for differences in emitter areas A_1 and A_2 and for resistors R_1 and R_2. Assume ideal transistors.

8. Use the schematic in Figure 3.6 to design a Widlar current mirror. $I_{ref} = 75\mu A$ and $R = 100\Omega$. What is I_{out}? What value of R is required for $I_{out} = 20\mu A$?

9. What is the output resistance of the circuit described in problem 1 if the Early voltage is 100V?

10. What is the output resistance of the Widlar mirror designed in exercise 1.8 if $V_A = 100V$?

11. Complete the small-signal analysis for the Wilson mirror and prove the validity of equation 3.18.

12. Design a V_{be}/R current source with $I_{out} = 25\mu A$. If the temperature coefficient of V_{be} is $-2mV/\deg C$ and the temperature coefficient of resistance is $+2500$ ppm, what is the tolerance on I_{out} from $-25\deg C$ to $+75\deg C$?

13. Design a $\Delta V_{be}/R$ current source with $I_{out} = 25\mu A$. If the temperature coefficient of V_{be} is $-2mV/\deg C$ and the temperature coefficient of resistance is $+2500$ ppm, what is the tolerance on I_{out} from $-25\deg C$ to $+75\deg C$?

14. Design a current mirror using n-channel MOS devices. Use the circuit shown in Figure 3.12. Assume supplies are $V_{DD} = 5V$ and ground. Reference current is $20\mu A$. Output currents are to be $20\mu A$, $40\mu A$, $55\mu A$, and $70\mu A$. $KP = 50\mu A/V^2$, $L = 5\mu m$, $V_{th} = 0.8V$ and $\Lambda = 0.04V^{-1}$. Determine the minimum voltage required to stay in saturation and then find the output resistance for each output.

15. Design a cascoded NMOS mirror using Figure 3.13 as a template. Use a reference current of $10\mu A$. Provide an output current of $50\mu A$ with an output resistance of $25M\Omega$. Use KP, V_{th} and Λ from exercise 14.

16. Redesign the cascoded mirror from exercise 15 to improve the output voltage swing. Use the circuit in Figure 3.15 as a template, and use the MOS device parameters from exercise 14.

References

[1] Baker, R. Jacob, et al, *CMOS Circuit Design, Layout and Simulation*, IEEE Press, New York, c. 1998.

[2] Gray, Paul R., and Mayer, Robert G., *Analysis and Design of Analog Integrated Circuits*, 2nd edition, John Wiley and Sons, Inc., New York, c. 1984.

[3] Millman, Jacob, and Grabel, Arvin, *Microelectronics*, 2nd edition, McGraw-Hill Book Company, New York, c. 1987.

chapter 4

Voltage References

4.1 Simple Voltage References

An ideal voltage reference produces a voltage that is constant and does not change with factors such as current loading, power supply variations, or temperature. Such references are useful in applications where a given voltage is compared to a standard such as in analog to digital converters and also for the generation of regulated supply voltages for digital circuits from a higher voltage analog supply.

A common way to provide a voltage is to use the voltage divider consisting of two resistors in series. The output voltage is $V_o = \frac{R_1}{R_1+R_2}V_{CC}$. The voltage divider has the advantage of simplicity, but the output is sensitive to power supply variations and to changes in current drawn from the V_o terminal.

The sensitivity of the output voltage to power supply variations is defined as

$$S_{V_{CC}}^{V_O} = \frac{\frac{dV_o}{V_o}}{\frac{dV_{CC}}{V_{CC}}} = \frac{V_{CC}}{V_o}\frac{dV_o}{dV_{CC}}$$

The fraction $\frac{dV_o}{V_o}$ can be expressed in units of percent change or in parts per million(ppm). For the voltage divider, the sensitivity to the power supply voltage is

$$S_{V_{CC}}^{V_O} = 1$$

This means the percent change in the output voltage equals the percent change in V_{CC}.

The sensitivity to load current is

$$S_{I_o}^{V_O} = \frac{I_o}{V_o}\frac{dV_o}{dI_o}$$

Since the output depends on the ratio of R_2 to R_1, it will be to a first order, independent of variations in resistance values. An accurate value for V_o depends on the ability to match resistance values.

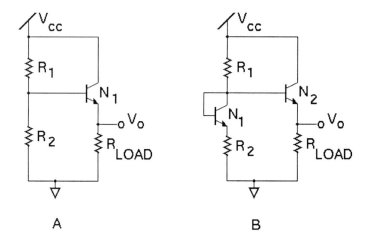

Figure 4.1   A. Buffered divider.   B. Temperature compensated buffered divider.

Consider the circuit shown in Figure 4.1A. The buffered divider shown uses a resistor divider to create a reference voltage. The NPN transistor provides relatively high levels of output current with minimal loading on the resistor divider node. The V_{be} of the transistor does lower the output reference voltage by about $0.7V$, and changing current levels will result in some change in the output voltage as the V_{be} changes. Additionally, supply voltage variation and temperature changes will affect the output voltage. But other than that, it's a great little circuit! We can get some temperature compensation by adding a second NPN to the circuit as shown in Figure 4.1B, but the other sensitivities are still there.

Since the buffered divider output is down one V_{be} from the divider reference voltage, temperature variations will cause the output to vary at approximately $2mV/\deg C$. The sensitivity to V_{cc} is unity, the same as the voltage divider.

4.2   Vbe Multiplier

The effects of supply variation can be minimized by "decoupling" the output from V_{cc}. Consider the circuit shown in Figure 4.2A. The base of transistor N_1 is at $3V_{be}$. The output V_o is $2V_{be} \approx 1.4V$. The output voltage is, to a first order, independent of V_{cc}. The output transistor N_1 can provide reasonable current, and changes in supply voltage affect the output voltage only insofar as the current through the diodes changes the diode V_{be} values. Doubling the value of V_{cc} will only raise the reference by about $54mV$. However, this circuit has its problems. It has a large negative temperature coefficient due to the multiple $V_{be}s$ in the reference diode string. Only multiples of V_{be} are possible as output reference voltage. Finally, the use of many diodes can consume large die areas, causing designs to grow and leading to higher manufacturing costs. Once again, we can address the shortcomings of this circuit. Figure 4.2B shows a circuit where we have used only two transistors, but note how the use of resistors R_1 and R_2 now allow the use of any multiple of V_{be} to be used.

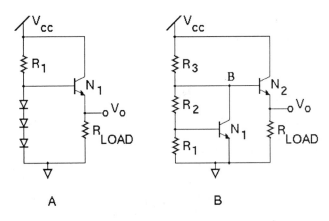

Figure 4.2   Vbe multiplier voltage references. V_o is a multiple of V_{be}. A. Diode dependent voltage reference. B. V_{be} multiplier voltage reference.

The configuration of these resistors and the transistor N_1 form a circuit known as a V_{be} multiplier. Neglecting base current to transistor N_1, the base-emitter voltage of N_1 is $\frac{R_1}{R_1+R_2}V_A$, where V_B is the voltage at the base of N_2. The output voltage V_o is one V_{be} below V_B:

$$V_o = \frac{R_1 + R_2}{R_1}V_{be} - V_{be} = \frac{R_1}{R_2}V_{be}$$

The current through R_3 has three components. R_3 provides current for the voltage divider R_1 R_2. It provides base current for N_2. The third component of current through R_3 is N_1 collector current. N_1 adjusts its collector current to assure V_B, the N_2 base voltage is correct provided R_3 is small enough to furnish sufficient current. The circuit will go out of regulation when the load current increases to a point where the base current to N_2 steals the current from the voltage divider that is required to maintain one V_{be} across R_1. N_1 acts as a feedback amplifier to stabilize the voltage at node B. As more current is drawn from the output, the voltage at node B, the base of N_1 drops. This causes the voltage across R_1 to drop. With its V_{be} decreasing, N_1 begins to turn off. With less current drawn from node B by N_1, the voltage at node B tends to increase and more current is available for base current to N_2.

The output voltage is to a first order, independent of V_{CC} and load current. It is however, sensitive to temperature, since V_{be} is temperature sensitive.

$$\frac{dV_o}{dT} = \frac{R_1}{R_2}\frac{dV_{be}}{dT} \approx \frac{R_1}{R_2}(-2mV/^oC)$$

Example

The circuit shown in Figure 4.2B has an output voltage of 5 V when $\frac{R_1}{R_2} = 7.7$. The output voltage decreases by about 15 mV per degree C. That's 0.3% per degree C or 3000 ppm/^oC.

4.3   Zener Voltage Reference

Figure 4.3A shows how a zener diode can be used to create a voltage reference. The zener diode is a structure that works based on avalanche breakdown. A large electric field across the base-emitter junction strips carriers away from lattice atoms, and the impact of these carriers on other atoms strips more carriers away, and so on. The result is current flow at a voltage much larger than V_{be}, often on the order of six to eight volts. The actual voltage depends on the doping levels and physical characteristics of the junction. Zener diodes usually have a positive temperature coefficient on the order of $+4\ mV/^oC$, and can be used to offset the temperature coefficient of V_{be}. The reference shown in Figure 4.3A is relatively independent of V_{cc}, and has reasonable temperature performance. It can only provide a maximum output current equal to the value of the current source less current equal to $\frac{V_{be}}{R}$. Note the position of Q1 with its emitter grounded. N_1 acts as a gain stage to keep the output voltage from going too high. Any increase in the output voltage causes the zener voltage to increase. This causes the zener to carry more current which drives the base of N_1 harder. Q1's collector voltage is then pulled down until equilibrium is restored. Figure 4.3B works in a manner analogous to the V_{be} multiplier. The multiplication of both V_z and V_{be} allows larger output voltage, but requires a much larger value of resistor for R_2 to keep the bias current small.

A few comments about zener diodes are appropriate here. Zeners are usually implemented as NPN transistors with the base-emitter junction reverse biased. The breakdown voltage of the base-emitter junction varies with the process. A typical value is $6.5V$. But zeners do have a problem associated with their use. Because zener current flows in the base-emitter junction, zener breakdown is primarily a surface phenomenon. The problem is that some of the highly energetic carriers flowing during zener breakdown become implanted in the oxide above

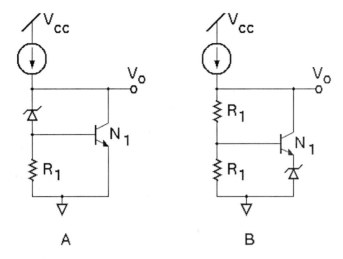

Figure 4.3   A. Temperature corrected zener reference.   B. High voltage reference.

the junction. This changes the electric field characteristics within the junction with the result that the zener voltage drifts as time passes. The change in zener voltage can be fairly large, often on the order of several hundreds of millivolts. The bottom line is that zeners aren't any good for developing precision references.

4.4   *Temperature Characteristics of I_c and V_{be}*

The current through an NPN transistor biased in the forward active region is given by

$$I_c(T) = I_s(T)e^{\frac{qV_{be}}{KT}}$$

where

$$I_s(T) = \frac{A_E q n_i^2(T) D_n(T)}{N_B}$$

If we evaluate this expression for I_c at two different temperatures, we can arrive at an accurate expression for $V_{be}(T)$. An arbitrary temperature T and a reference temperature T_r are chosen. The result of some algebraic manipulation is

$$V_{be} = V_g(T) - \frac{T}{T_r}V_G(T_r) + \frac{T}{T_r}V_{be}(T_r) - \eta\frac{KT}{q}ln\left[\frac{T}{T_r}\right] + \frac{KT}{q}ln\left[\frac{I_c(T)}{I_c(T_r)}\right]$$

$V_G(T)$ is the bandgap voltage of silicon, which is a non-linear function as temperature decreases. However, replacing Vg(T) with Vg0, the linear extrapolation of V_{be} at $0\,^oK$ is a good approximation for temperatures of interest above 200^oK (-70°C). The term η represents the

temperature dependence of carrier mobility in silicon and is equal to
$4 - n$, where n is taken from

$$\mu(T) = CT^{-\eta} \tag{4.1}$$

From past history, designers know that forcing the collector current in
the transistor to be proportional to absolute temperature (PTAT) helps
to reduce the effect of η. If we make this assumption, then the equation
for V_{be} can be simplified to

$$V_{be}(T) = V_G(T) + \frac{T}{T_r}[V_{be}(T_r) - V_G(T_r)] - (\eta - 1)\frac{KT}{q}ln\left[\frac{T}{T_r}\right]$$

Depending on the magnitude of the collector current, we know the
change in V_{be} due to temperature is about -2 mV per degree Centigrade. We would like to balance this temperature variation by adding
a voltage that has a positive temperature coefficient in order to obtain
a temperature-invariant reference voltage. We know that the thermal
voltage V_T is proportional to absolute temperature, and we can develop
our reference using some multiple of V_T to cancel the V_{be} temperature
coefficient.

Using the magnitudes shown in Figure 4.4, we can calculate that these
variations exactly offset each other if we have

$$\Delta V_{be} + K(\Delta V_T) = 0$$

This is the case if we take $K = 20.9$. Then we have

$$V_{ref} = V_G(T) + \frac{T}{T_r}[V_{be}(T_r) - V_G(T_r)] - (\eta - 1)\frac{KT}{q}ln\left[\frac{T}{T_r} + K(V_T)\right]$$

The above equation for V_{ref} is plotted in Figure 4.4 as a function of
frequency and η. It shows a voltage variation of only a few millivolts
over a wide temperature range.

Since we have made some approximations in this derivation, it is useful
to observe the following:

- Some non-linearity is present in V_{ref} due to the effects of η and
  due to changes in $V_G(T)$, especially as T drops below about -60°C.

- The lowest theoretical temperature coefficient for V_{ref} is about 15
  parts per million per degree Centigrade, depending on the value
  of η.

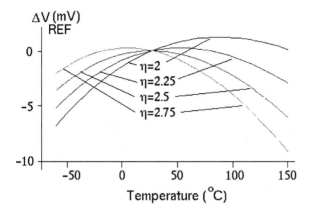

Figure 4.4   Simulation showing temperature variation and η dependence of the bandgap voltage V_{ref}. η describes the temperature variation of mobility. (See Equation 4.1)

4.5   Bandgap Voltage Reference

The bandgap circuit shown in Figure 4.1 produces a voltage V_{bg} that is, to a first order, temperature and supply independent and approximately equal to the silicon bandgap voltage of 1.2 V. The voltage divider formed by R_4 and R_5 multiplies V_{bg} to produce higher voltages at V_o. The current mirror P_1, P_2, acts to hold $I_1 = I_2$.

$$I_1 = nI_s e^{\frac{V_{be_1}}{V_T}} = I_2 = I_s e^{\frac{V_{be_2}}{V_T}}$$

$$V_T \ln[n] = V_{be_2} - V_{be_1} = R_2 I_1$$

solving for $I = I_1$

$$I = \frac{V_T \ln[n]}{R_2} \tag{4.2}$$

The voltage at the base of N_2 is the bandgap voltage

$$V_{bg} = R_1 2I + V_{be_2}$$

Using Equation 4.2

$$V_{bg} = R_1 2 \frac{V_T \ln[n]}{R_2} + V_{be_2} \tag{4.3}$$

where $V_T = \frac{KT}{q} = 0.026V$ at $T = 300^\circ K$. V_T is proportional to the absolute temperature, $V_T = 8.62x10^{-5}T$.

V_T increases with temperature while V_{be} decreases with temperature at about -2mV per degree C. The first term in Equation 4.3 increases

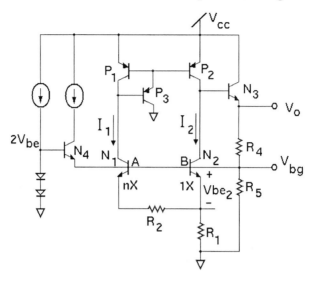

Figure 4.5   Bandgap voltage reference circuit produces a voltage insensitive to temperature and supply voltage.

with temperature and the second term decreases with temperature. When these changes are made to compensate each other, changes with temperature are minimized. Taking the derivative of Equation 4.3 with respect to temperature and setting it equal to zero and rearranging terms

$$2\frac{R_1}{R_2}\ln[n] = \frac{0.002}{8.62x10^{-5}}$$

$$\frac{R_1}{R_2} = \frac{11.6}{\ln[n]}$$

If $n = 4$, $\frac{R_1}{R_2} = 8.4$.

Example

For the bandgap circuit in Figure 4.5, if $n = 4$ the voltage across R_2 is about $36mV$. If R_2 equals 450 ohms, I will equal $80\mu A$ and R_1 should equal about 3.7K. The drop across R_1 is $2IR_1 = 2x10x10^{-6}x3.7x10^3 = 0.6V$.

The bandgap voltage V_{bg} is $0.6 + V_{be_2} = 0.6 + 0.65 = 1.25V$.

Feedback Mechanism

Transistors N_1, N_2, P_1, and P_2 form an amplifier. These transistors, together with transistor N_3 provide a feedback signal that stabilizes the bandgap voltage. Consider N_1 and N_2 redrawn in Figure 4.3A. Recall that collector current depends on base emitter voltage.

$$I_c = nI_s e^{\frac{V_{be}}{V_T}}$$

where n is the emitter multiplication factor.

At low currents the V_{be}s of the two transistors are nearly equal because the drop across R is small. As shown in Figure 4.6, with nearly equal V_{be}s, I_1 is greater than I_2 because N_1 is larger then N_2. As the base voltages increase, currents increase. The current in N_1 is limited by R to approximately linear increases, while the current in N_2 increases exponentially with V_{be_2}.

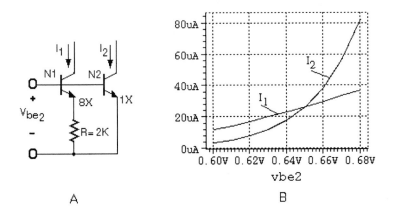

A   B

Figure 4.6   There is an input voltage at which the currents are equal (about 0.65V in this simulation). If V_{be_2} increases above that value, $I_2 > I_1$. If it drops below it, $I_2 < I_1$.

References

[1] A. Paul Brokaw, *A Simple Three-Terminal IC Bandgap Reference*, IEEE Journal of Solid State Circuits, Volume SC-9, No. 6, December 1974.

[2] P.R. Gray and R.G. Meyer, *Analysis and Design of Analog Integrated Circuits*, 2nd edition, Wiley, New York, c. 1984, pp. 233-246, 289-296.

[3] Brian Harnedy, *ELE536 Class Notes: Circuit 513: A Bandgap Referenced Regulator*, Cherry Semiconductor Memorandum, 1987.

[4] C. Tuozzolo, *Voltage References and Temperature Compensation*, Cherry Semiconductor Memorandum, 1996.

chapter 5

Amplifiers

Bipolar and MOSFET transistors are both capable of providing signal amplification. There are three amplifier types that can be obtained using a single transistor. These amplifiers are described in the chart below.

Bipolar Technology

Configuration	Signal Applied To	Output Taken From
Common-emitter	Base	Collector
Common-base	Emitter	Collector
Common-collector	Base	Emitter

MOS technology

Configuration	Signal Applied To	Output Taken From
Common-source	Gate	Drain
Common-gate	Source	Drain
Common-drain	Gate	Source

The common-collector amplifier and the common-drain amplifier are often referred to as the emitter follower and the source follower, respectively.

There are several frequently used two-transistor amplifiers to be considered as well. These are the Darlington configuration, the CMOS inverter, the cascode configuration and the emitter-coupled (or source-coupled) pair. The cascode amplifier and the coupled-pair amplifier are available in both bipolar and MOS technologies.

Each of these amplifier types will have its own characteristics: voltage and current gain, and input and output resistance. Analysis of complicated circuits can be simplified by considering the large circuit as a combination of simpler blocks.

In this chapter, we will present the bipolar case first and then repeat our analyses for the MOS equivalent circuits.

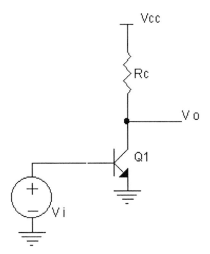

Figure 5.1 Common-emitter amplifier.

5.1 The Common-Emitter Amplifier

The schematic for the resistor-loaded common-emitter amplifier is shown in Figure 5.1. The circuit load is shown as resistor R_C. Let us start by evaluating the amplifier's transfer function as the value of the input source V_I is increased.

With $V_I = 0$, transistor Q1 is cut off. There is no current flow in the base, so collector current is also zero. Without current in the collector, there is no voltage developed across R_C, and $V_o = VCC$. As V_I increases, Q1 enters the forward active region and begins to conduct current. Collector current can be calculated from the diode equation:

$$I_C = I_S \exp\left(\frac{V_I}{V_T}\right) \tag{5.1}$$

The large-signal equivalent circuit is provided below. As the value of V_I increases, there is an exponential gain in collector current. As collector current increases, the voltage drop across R_C also increases until Q1 enters saturation. At this point, the collector to emitter voltage of Q1 has reached its lower limit. Further increase of the input voltage will provide only very small changes in the output voltage.

The output voltage is equal to the supply voltage minus the drop across the collector resistor:

$$V_o = V_{CC} - I_C R_C = V_{CC} - R_C I_S \exp\left(\frac{V_I}{V_T}\right) \tag{5.2}$$

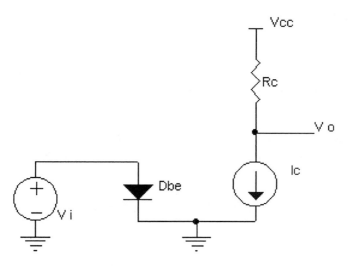

Figure 5.2   Common-emitter amplifier large signal equivalent circuit.

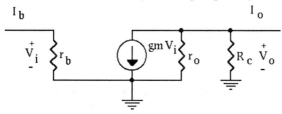

Figure 5.3   Common-emitter amplifier small signal equivalent circuit.

Plotting the transfer function shows an important result. A small incremental change in V_I causes a large change in V_o while Q1 operates in forward active mode. The circuit exhibits voltage gain.

We can use the small-signal equivalent circuit shown in Figure 5.3 to calculate the gain. In this analysis, we do not include high frequency model components. We also ignore the internal resistance of the source V_I and the resistance of any load driven from V_o.

The small signal analysis gives

$$V_o = -g_m V_I (r_o \parallel R_C) \tag{5.3}$$

The unloaded voltage gain is then given as

$$A_V = \frac{V_o}{V_I} = -g_m (r_o \parallel R_C) \tag{5.4}$$

The input resistance is given as

$$R_I = r_b \tag{5.5}$$

and the output resistance is

$$R_o = r_o \parallel R_C \qquad (5.6)$$

If the value of R_C is allowed to approach infinity, the gain equation for the common-emitter amplifier reduces to

$$A_V = -g_m r_o = -\frac{V_A}{V_T} \qquad (5.7)$$

Finally, we can calculate the short circuit current gain. If we short the output, we obtain

$$I = \frac{I_o}{I_I} = \frac{g_m V_I}{\frac{V_I}{r_b}} = g_m r_b = \beta \qquad (5.8)$$

Example

Use the circuit defined in Figure 5.1 with $R_C = 10\ K\Omega$, $I_C = 50\ \mu A$, $beta = 100$, and $r_o = \infty$. Find input resistance, output resistance, voltage gain and current gain for the common-emitter amplifier.

Solution

Input resistance: $R_I = r_b = \beta/g_m = 100/(50\ \mu A/26\ mV) = 52\ K\Omega$
Output resistance: $R_o = r_o \parallel R_C \approx R_C = 10\ K\Omega$
Voltage gain: $A_V = -g_m R_o = -(50\ \mu A/26\ mV)/10\ K\Omega = -19.23$
Current gain: $A_I = \beta = 100$

Our analysis to this point has ignored external loading effects. Let us add some base resistance and load resistance to our circuit and find the effects on voltage gain.

We will start our analysis by assuming that V_I's DC level is adjusted to maintain $I_C = 50\ \mu A$. Let $R_b = 10\ K\Omega$ and $R_L = 10\ K\Omega$. The small signal equivalent circuit is shown in Figure 5.5.

From this we have

$$V_I = V_S \frac{r_b}{R_b + r_b}$$

and

$$V_o = -g_m V_I (R_o \parallel R_L) = -g_m V_S \frac{r_b}{R_b + r_b} \frac{R_o R_L}{R_o + R_L}$$

so that

$$A_V = \frac{V_o}{V_S} = -g_m \frac{r_b}{R_b + r_b} \frac{R_o R_l}{R_o + R_L}$$

$$A_V = -\left(\frac{50\ \mu A}{26\ mV}\right)\left(\frac{52\ K\Omega}{62\ K\Omega}\right) 5\ K\Omega = -8.065$$

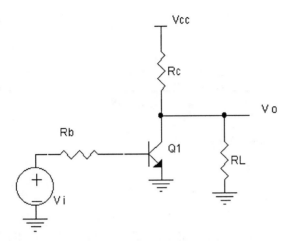

Figure 5.4   Resistor loaded common-emitter amplifier.

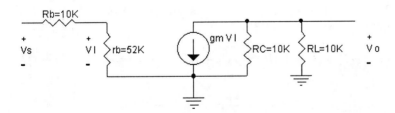

Figure 5.5   Small signal equivalent circuit for the resistor loaded common-emitter amplifier.

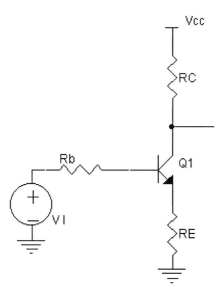

Figure 5.6 Common-emitter amplifier with emitter degeneration resistor, R_E.

Another circuit option for the common-emitter amplifier is shown in Figure 5.6. Here we see the addition of a series resistance between the emitter and ac ground. The presence of this resistance increases output resistance, increases input resistance and decreases transconductance. The resulting decrease in voltage gain leads us to call the presence of this resistance emitter degeneration. The equivalent circuit in Figure 5.7 will be used to determine input resistance and transconductance while Figure 5.8 will help us calculate output resistance.

Let us look at input resistance assuming $r_o \to \infty$ and $R_b = 0$. From Figure 5.7, we see that

$$V_I = r_b I_b + (I_b + I_c)R_E = r_b I_b + I_b(\beta + 1)R_E = I_b(r_b + (\beta + 1)R_E)$$

$$R_I = \frac{V_I}{I_b} = r_b + (\beta + 1)R_E$$

If β is large, we can say that $R_I \approx r_b + \beta R_E$, and since $\beta = g_m r_b$, we have $R_I \approx r_b(1 + g_m R_E)$.

Considering transconductance, Figure 5.7 again shows that

$$V_I = r_b I_b + (I_b + I_c)R_E = \frac{I_c}{\beta}r_b + I_c\left(1 + \frac{1}{\beta}\right)R_E = I_c\left(\frac{1}{g_m} + R_E + \frac{R_E}{\beta}\right)$$

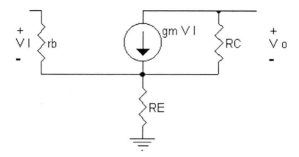

Figure 5.7   Small signal equivalent circuit for the common-emitter amplifier with emitter degeneration.

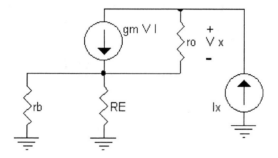

Figure 5.8   The test current I_x is used to calculate the output resistance.

Amplifier transconductance is then

$$G_m = \frac{I_c}{V_I} = \frac{1}{\frac{1}{g_m} + R_E + \frac{R_E}{\beta}}$$

If β is large

$$G_M \approx \frac{1}{\frac{1}{g_m} + R_E} = \frac{g_m}{1 + g_m R_E}$$

Output resistance is determined by using a test current and calculating the resulting voltage.

We first assume that R_C is very large and can be ignored. Next we note that the entire test current flows in the parallel combination of r_b and R_E. This gives

$$V_I = I_X (r_b \parallel R_E)$$

We also note that current flowing through r_o is given by

$$I(r_o) = I_x - g_m V_I = I_x + I_x g_m (r_b \parallel R_E)$$

Using these results we find voltage V_x

$$V_x = -V_I + I(r_o)r_o = I_x[r_b \parallel R_E + r_o(1 + g_m(r_b \parallel R_E))]$$

Finally, we have $R_o = V_x/I_x$ such that

$$R_o = r_b \parallel R_E + r_o(1 + g_m(r_b \parallel R_E))$$

The second term is much larger than the first, so we can neglect the first to obtain

$$R_o = r_o(1 + g_m(r_b \parallel R_E)) = r_o\left(1 + g_m\frac{r_b R_E}{r_b + R_E}\right)$$

If we divide both the numerator and denominator of the fractional term by r_b and use the identity $r_b = \beta g_m$, we arrive at

$$R_o = r_o\left(1 + \frac{g_m R_E}{1 + \frac{g_m R_E}{\beta}}\right)$$

If β is much larger than $g_m R_E$, this reduces to

$$R_o = r_o(1 + g_m R_E)$$

Finally, we can evaluate the voltage gain of the degenerated common-emitter amplifier using these simplifying assumptions, but also assuming a finite value of R_C:

$$A_V = -G_m(R_o \parallel R_C) = -\frac{g_m}{1 + g_m R_E}\frac{r_o(1 + g_m R_E)R_C}{r_o(1 + g_m R_E) + R_C} = -\frac{g_m R_C}{1 + g_m R_E + \frac{R_C}{r_o}}$$

If R_C/R_o is small compared to $(1 + g_m R_E)$, the voltage gain reduces to

$$A_V \approx -\frac{R_C}{R_E}$$

This is a very important result. If all our assumptions are valid, we can design amplifiers whose gain is independent of g_m and β variations.

5.2   The Common-Base Amplifier

The common-base amplifier has a signal applied to the emitter and the output is taken from the collector. The base is tied to ac ground. This circuit is frequently used in integrated circuits to increase collector resistance in current sources. This technique is called cascoding.

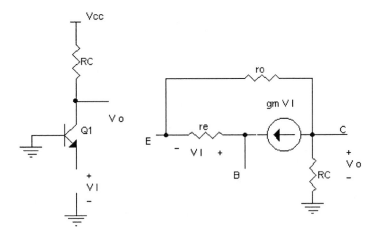

Figure 5.9   Common-base amplifier and simplified "T-model."

The hybrid-*pi* model is an accurate tool, but it is difficult to use for this circuit. Gray and Meyer suggest a simplified "T-model" that is easy to use and understand, although it is limited to low frequency cases where R_C is much smaller than r_o of the transistor. The circuit schematic and simplified "T-model" are shown in Figure 5.9. Note that r_o should be ignored unless $R_C \approx r_o$, at which time r_o should be included in the analysis.

The simplification process results in the creation of a new circuit element r_e. This resistance is the parallel combination of r_b and a controlled current source modeled as a resistance of value $1/g_m$. Thus,

$$r_e = \frac{1}{g_m + \frac{1}{r_b}} = \frac{1}{g_m(1 + \frac{1}{\beta})} = \frac{\beta}{g_m(\beta + 1)} = \frac{\alpha}{g_m}$$

If β is large, $r_e \approx V_T/I_C$.

By inspection, the input resistance is seen to be $R_I = r_e$. Output resistance is similarly $R_o = R_C$. Transconductance is $G_m = g_m$. From this, we find the voltage gain and current gain:

$$A_V = G_m R_o = g_m R_C$$

$$A_I = G_m R_I = g_m R_e = \alpha$$

Note that the current gain of this amplifier topology is always less than unity. This makes the cascading of several amplifiers impractical without some type of gain stage between the common-base stages.

In addition to cascoding, common-base amplifiers are not subject to high-frequency feedback from output to input through the collector-base capacitance as are common-emitter amplifiers.

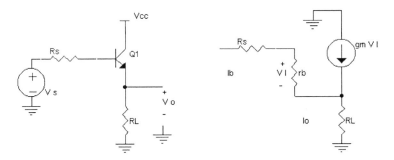

Figure 5.10   Emitter follower and small signal equivalent circuit.

5.3   Common-Collector Amplifiers (Emitter Followers)

The common-collector amplifier has its input signal applied to the base and the output is taken at the emitter. In this circuit, input resistance depends on the load resistance and output resistance depends on the source resistance. The circuit schematic and small-signal equivalent circuit are shown in Figure 5.10.

By summing currents at the output node, we can find the voltage gain V_o/V_s:

$$A_V = \frac{V_o}{V_s} = \frac{1}{1 + \frac{R_s + r_b}{(\beta + 1)R_L}}$$

If we replace the voltage source and R_s with a test current source

$$R_I = \frac{V_x}{I_x} = r_b + R_E(\beta + 1)$$

The input resistance is increased by $\beta + 1$ times the emitter resistance. Replacing R_L with a test voltage source allows us to determine the output resistance:

$$R_o = \frac{V_x}{I_x} \approx \frac{1}{g_m} + \frac{R_s}{\beta + 1}$$

The output resistance equals the base resistance divided by $\beta + 1$, plus $1/g_m$.

One of the major uses of emitter followers is as an "impedance matcher." It has high input resistance, low output resistance, voltage gain near unity and can provide current gain. The emitter follower is often placed between an amplifier output and a low impedance load. This can help reduce loading effects and keep amplifier stage gain high.

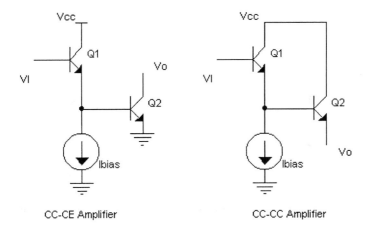

CC-CE Amplifier   CC-CC Amplifier

Figure 5.11 Common-emitter and common-collector two-transistor amplifiers.

5.4 Two-Transistor Amplifiers

Typical one-transistor amplifiers can provide voltage gain of several thousand depending on loading. Thus, most practical IC amplifiers require several stages of amplification to provide the levels of performance needed for circuit applications of today. Analysis of cascaded stages can be completed by considering each transistor as a stage, but several widely used two-transistor "cells" exist. These can be considered single stages and analysis can be simplified. Five common subcircuits will be discussed here: the common-collector, common-emitter amplifier (CC-CE); the common-collector, common-collector amplifier (CC-CC); the Darlington configuration; the common-emitter, common-base amplifier, also known as the cascode; and the common-collector, common-base amplifier, also known as the emitter-coupled pair.

MOS equivalents to the cascode and emitter-coupled pair circuits exist, but analogues for the CC-CE, CC-CC and Darlington configurations can be better implemented as physically larger single transistor designs.

5.5 CC-CE and CC-CC Amplifiers

CC-CE and CC-CC configurations are shown in Figure 5.11. Note that some type of bias element is usually required to set the quiescent operating point of transistor Q1. The element may be a current source, a resistor, or it may be absent. Q1 is present for two main reasons. It increases the current gain of the stage, and it increases the input resistance of the stage. These circuits can be considered as a single "composite" transistor as long as Q1's output resistance doesn't affect the circuit

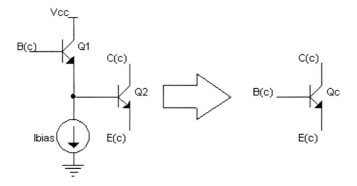

Figure 5.12 Two-transistor amplifiers can be represented by one composite transistor.

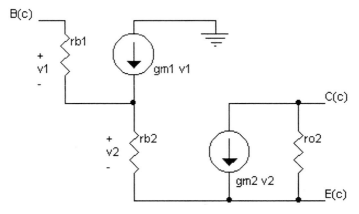

Figure 5.13 Small signal equivalent circuit for the composite transistor.

performance. Let us determine the circuit values of r_b, r_o, g_m, and β for the composite transistors. We'll denote the values for the composite transistor with a suffix (c) as shown in Figure 5.12.

The small signal equivalent circuit for the composite transistor is shown in Figure 5.13. We assume the effects of r_{o1} are negligible for this analysis.

We first look at input resistance with the composite emitter grounded. The input resistance to Q2 is simply r_{b2}. Thus, the input resistance to the composite transistor looks like the input resistance to the common-emitter amplifier with emitter degeneration. In this case

$$r_b(c) = r_{b1} + (\beta + 1)r_{b2}$$

Next, we consider transconductance. Transconductance of the composite transistor is the change in collector current of Q2 for a given change in the effective v_{be} of the composite. We need to know how v_2 changes for a change in the composite v_{be}. This circuit looks like the emitter follower found in the previous section:

$$\frac{v_2}{v_{be}} = \frac{1}{1 + \frac{r_{b1}}{(\beta+1)R_{b2}}}$$

Since collector current of the composite is identical to the current of Q2, we have

$$I_C(c) = g_m(c)v_{be}(c) = g_{m2}v_2 = \frac{g_{m2}V_{be}(c)}{\frac{r_{b1}}{(\beta+1)r_{b2}}}$$

and so the composite transconductance is

$$g_m(c) = \frac{I_C(c)}{v_{be}(c)} = \frac{g_{m2}}{1 + \frac{r_{b1}}{(\beta+1)r_{b2}}}$$

In the case where $I_{bias} = 0$, emitter current of Q1 is equal to base current of Q2. This results in $r_{b1} = \beta r_{b2}$, and transconductance of the composite simplifies to

$$g_m(c) = \frac{g_{m2}}{2}$$

Current gain of the composite is the ratio of $I_C(Q2)$ to $I_B(Q1)$. The base current of Q2 is equal to the emitter current of Q1 such that

$$I_C(c) = I_C(Q2) = \beta I_E(Q1) = \beta(\beta+1)I_B(Q1)$$

so that

$$\beta(c) = \beta(\beta+1) \approx \beta^2$$

Finally, by inspection we have $r_o = r_{o2}$.

5.6   The Darlington Configuration

The Darlington configuration is shown in Figure 5.14. It is characterized by having the collectors tied together, and the emitter of one transistor drives the base of the other. This composite can be used in common-emitter, common-collector or common-base configurations. It is usually good design practice to have some type of bias element present for Q1, but that element is not required. When used as a common-emitter amplifier, the collector of Q1 connects to the output instead of to an ac ground, and it provides feedback that reduces $r_o(c)$. Input capacitance is also increased because the collector-base capacitance of Q1 is connected between the input and output, resulting in Miller-effect multiplication. These drawbacks often make it preferable to use the CC-CE configuration.

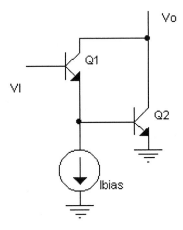

Figure 5.14   Darlington configuration.

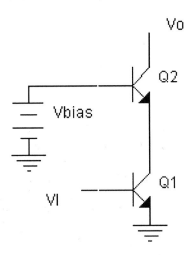

Figure 5.15   Cascode amplifier.

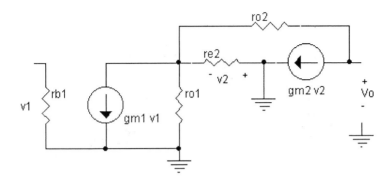

Figure 5.16   Small signal equivalent circuit for the cascode amplifier.

5.7   *The CE-CB Amplifier, or Cascode*

The cascode amplifier is shown in Figure 5.15. We have seen how cascoding increases output resistance in current mirrors. This is again an advantage in amplification stages, since voltage gain typically resembles $g_m r_o$. Another advantage is that there is no high-frequency feedback path through the collector-base capacitance as occurs in the common-emitter topology.

The small signal equivalent circuit for the cascode is developed by taking the equivalent circuits for both CE and CB stages and putting them together. This circuit is shown in Figure 5.16. We will use the equivalent circuit to determine input resistance, output resistance and transconductance.

We see by inspection that the input resistance of the cascode is simply r_b of Q1. We also see that the current gain from emitter to collector of Q2 is approximately one, so the transconductance of the cascode is approximately the transconductance of Q1, or g_{m1}.

To calculate output resistance, we first simplify our equivalent circuit. Shorting v_1 to ground has the result of removing r_{b1} and the controlled current source for Q1. We also note that r_{o1} and r_{e2} are now in parallel to ground. Since r_{o1} is much larger than r_{e2}, we can neglect r_{o1} as well. We can redraw our equivalent circuit and connect a test voltage to our output as shown in Figure 5.17.

Current I_{x1} is given by

$$I_{x1} = \frac{V_x}{r_{o2} + r_{e2}} \approx \frac{V_x}{r_{o2}}$$

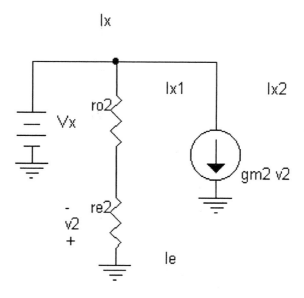

Figure 5.17 Redrawn small signal equivalent circuit for determining the output impedance.

This current is also equal to $-I_e$. Current I_{x2} flows in the controlled current source, but this current is defined as

$$I_{x2} = g_{m2}v_2 = \alpha I_e = -\alpha \frac{V_x}{r_{o2}}$$

Total current $I_x = I_{x1} + I_{x2}$ such that

$$I_x = \frac{V_x}{r_{o2}}(1 - \alpha) \approx \frac{V_x}{\beta r_{o2}}$$

and

$$R_o = \beta r_{o2}$$

5.8 Emitter-Coupled Pairs

Emitter-coupled pairs, also known as differential amplifiers, are probably the most often used type of amplifier in integrated circuit design. The emitter-coupled pair provides differential input characteristics required for all operational amplifiers. Cascading of sequential stages can be accomplished without need for impedance matching, and relatively high gains can be realized in a small area of circuitry, especially when combined with current mirror active loads. The MOS differential amplifier is called a source-coupled pair.

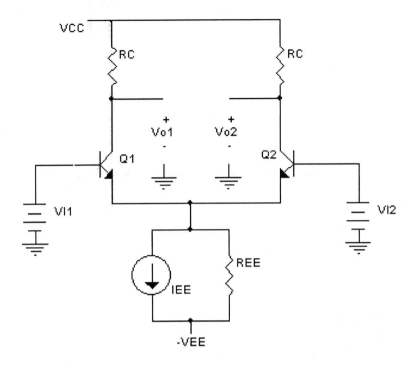

Figure 5.18   Emitter-coupled pair.

The schematic representation of the emitter-coupled pair is shown in Figure 5.18. Note that the biasing current source can be a transistor source (current mirror) or a simple resistor. If a resistor is used, the current source I_{EE} becomes zero. If a transistor is used, the transistor equivalent circuit replaces I_{EE} and the resistor.

Let us first consider the large signal transfer characteristic of the emitter-coupled pair shown in Figure 5.18. For simplicity, we will assume the bias current source output resistance and the output resistances of Q1 and Q2 are all infinite. This assumption is valid for the large signal analysis, but not for the small signal analysis. We can also assume that Q1 and Q2 are identical transistors with the same saturation current I_S. If we use Kirchoff's Voltage Law on the loop containing V_{I1}, V_{I2} and the base-emitter junctions of Q1 and Q2, we obtain

$$V_{I1} - V_{be1} + V_{be2} - V_{I2} = 0$$

Recalling that $V_{be} = V_T \ln[I_C/I_S]$, we can rewrite the equation above and solve for the ratio of collector currents I_{C1} and I_{C2}:

$$\frac{I_{C1}}{I_{C2}} = \exp\left(\frac{V_{I1} - V_{I2}}{V_T}\right)$$

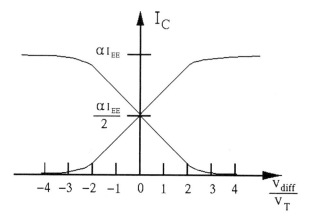

Figure 5.19   Collector currents as a function of the input voltage.

Next we sum currents at the node where the emitters of Q1 and Q2 are connected:

$$-(I_{E1} + I_{E2}) = \frac{I_{C1}}{\alpha} + \frac{I_{C2}}{\alpha} = I_{EE}$$

We now solve for the collector currents

$$I_{C1} = \frac{\alpha I_{EE}}{1 + \exp\left(-\frac{V_{diff}}{V_T}\right)}$$

and

$$I_{C2} = \frac{\alpha I_{EE}}{1 + \exp\left(\frac{V_{diff}}{V_T}\right)}$$

These currents are plotted as a function of V_{diff} in Figure 5.19. Note that the currents become independent of V_{diff} for values greater than $3V_T$, or about 75 mV. At this point, all of the current I_{EE} is flowing in only one of the transistors. The current change is linear for a region slightly less than about $\pm 2V_T$.

The output voltages are given as

$$V_{o1} = V_{CC} - I_{C1}R_C$$

and

$$V_{o2} = V_{CC} - I_{C2}R_C$$

However, it is often the case that the differential output voltage, $V_{odiff} = V_{o1} - V_{o2}$, is of most interest. V_{odiff} is plotted against V_{diff} in Figure 5.20.

It is possible to extend the range of linear operation by the addition of emitter degeneration resistors as shown in Figure 5.21. The linear region

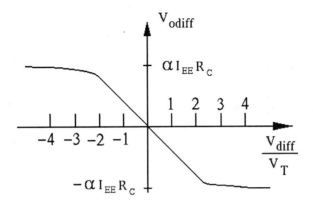

Figure 5.20   Differential pair output voltage as a function of the differential input voltage.

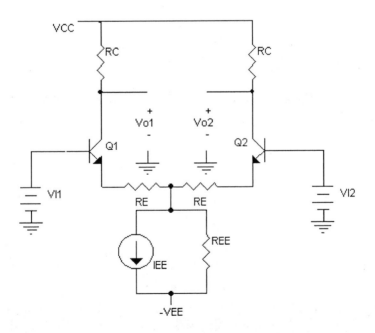

Figure 5.21   The emitter degeneration resistors, R_E, extend the linear range of the emitter-coupled pair.

is extended by about $\pm I_{EE} R_E$, but the voltage gain will be decreased as a result of adding degeneration.

For emitter-coupled pairs, we are most often interested in the small signal analysis when the dc differential input voltage is zero. In this case, V_{diff} represents the ac signal. In analyzing this circuit, we make the following assumptions:

- The magnitude of the input signal V_{diff} is small enough that the amplifier operates in the linear region.

- The equivalent resistance of the biasing circuitry is finite.

- r_o for the transistors is much larger than R_C and can be ignored in our analysis.

It is convenient to define the input signal as a sum of two components, a dc common-mode voltage and an ac differential-mode voltage. The differential-mode signal is defined as the difference between the two inputs, while the common-mode signal is the average of the two inputs. That is

$$V_{id} = V_{I1} - V_{I2} = V_{diff}$$

and

$$V_{ic} = \frac{V_{I1} + V_{I2}}{2}$$

We can redraw our circuit in Figure 5.22 to see the significance of these definitions.

We can similarly define differential-mode and common-mode output signals

$$v_{od} = v_{o1} - v_{o2}$$

and

$$v_{oc} = \frac{v_{o1} - v_{o2}}{2}$$

We can identify v_{o1} and v_{o2} in terms of v_{od} and v_{oc}

$$v_{o1} = \frac{v_{od}}{2} + v_{oc}$$

and

$$v_{o2} = -\frac{v_{od}}{2} + v_{oc}$$

The differential-mode gain is the change in the differential-mode output for a unit change in differential-mode input. Common-mode gain is similarly the change in common-mode output for a change in the common-mode input

$$A_d = \frac{v_{o1} - v_{o2}}{v_{i1} - V_{i2}}$$

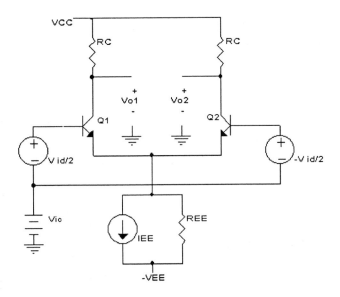

Figure 5.22 The input voltages can be represented in terms of a differential voltage V_{id}, and a common-mode voltage V_{ic}.

and

$$A_c = \frac{v_{o1} + v_{o2}}{v_{i1} + V_{i2}}$$

In order to complete our analyses, we return to the circuit in Figure 5.22, and set the common-mode supply to zero. This gives us a purely differential-mode circuit. Further, if we consider the circuit operation, we can see that the emitter connection serves as an ac ground. No ac current flows in R_{EE}. We can therefore reduce the emitter-coupled pair to the small signal equivalent shown in Figure 5.23A, as the circuit is completely symmetrical. Because of this, we can analyze the entire circuit by considering only one side of it. We can then further reduce the equivalent circuit to the one shown in Figure 5.23B. This reduced equivalent is called the differential half-circuit.

A quick analysis shows that the differential-mode gain is $A_d = v_{od}/v_{id} = -g_m R_C$. Let us now consider the circuit in Figure 5.22 with the differential voltages set to zero. This results in a purely common-mode input. The small signal equivalent for this circuit is shown in Figure 5.24A.

Note that we have replaced the resistor R_{EE} with two parallel resistances of $2R_{EE}$. The total resistance from the emitters to ground has not changed. Note also that the same voltage is applied to both bases, and $V_1 = V_2$. This means the collectors are conducting the same currents. This also implies no current is flowing in the connection between the two emitters. We can then remove that connection as shown in Fig-

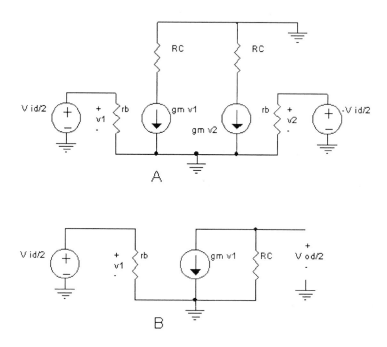

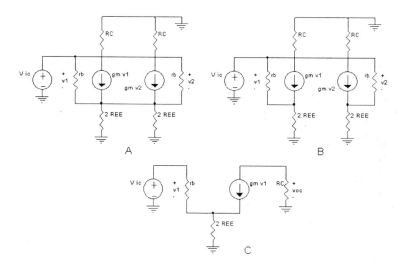

Figure 5.23   The small signal equivalent circuit shown in A can be reduced to the differential-mode half-circuit shown in B.

Figure 5.24   The small signal equivalent circuit shown in A can be reduced to the common-mode half-circuit shown in C.

ure 5.24B. Again, we have a symmetrical circuit that can be analyzed by the half-circuit concept. The common-mode half-circuit is shown in Figure 5.24C.

If we use Kirchoff's Voltage Law around the loop containing the input source r_b and the emitter resistance, we can solve for the base current

$$I_b = \frac{V_{ic}}{r_b + 2R_{EE}\left(1 + \frac{1}{\beta}\right)}$$

The common-mode output voltage is $-R_C\beta I_b$, and the common-mode voltage gain is then

$$A_c = -\frac{g_m R_C}{1 + 2g_m R_{EE}\left(1 + \frac{1}{\beta}\right)}$$

Note that if β is large and if $2g_m R_{EE}$ is much larger than unity, common-mode gain reduces to

$$A_c \approx \frac{R_C}{2R_{EE}}$$

The study of common-mode amplifier operation does not explain the physical consequences of changing common-mode input voltage. As the common-mode input voltage changes, the voltage across R_{EE} will change, since the transistor V_{BE}s will remain approximately constant. This results in a change in collector current and a shift in the common-mode output voltage.

Ideally, differential gain is high while common-mode gain is zero. We can get a feel for how close our circuits are to the ideal by evaluating the common-mode rejection ratio, or $CMRR$:

$$CMRR = \frac{A_d}{A_c} = 1 + 2g_m R_{EE}\left(1 + \frac{1}{\beta}\right)$$

Increasing the resistance R_{EE} of the current source decreases A_C and increases $CMRR$.

Differential input resistance is defined as the ratio of differential input voltage to small signal base current. That is

$$R_{id} = \frac{v_{id}}{I_b}$$

but $I_b = \frac{v_{id}}{2}/r_b$, so

$$R_{id} = 2r_b$$

Differential input resistance is dependent on r_b which increases with β and decreases with collector current. High differential input resistance requires operating the emitter-coupled pair at low collector currents. Common-mode input range is defined as the range of common-mode input voltage over which the amplifier can operate in the linear region. The main constraints on this range tend to be voltage requirements to keep the emitter-coupled pair out of saturation. For example, consider again the circuit in Figure 5.22. For a given current I_{EE}, a certain finite voltage is required across R_{EE} whether the bias element is a resistor or a transistor current mirror. Additionally, the V_{BE}s of Q1 and Q2 must be large enough for the bias currents to flow in the transistors. The minimum value of the common-mode input voltage must provide for these conditions to exist. Similarly, the voltage dropped across the collector resistances is given as I_C times R_C. If we define a voltage $V_C = V_{CC} - I_C R_C$, then raising v_{ic} above V_C will result in the transistor being pushed into the saturation region. The maximum value of the common-mode input range is then approximately V_C.

Input offset voltage is defined as the differential input voltage required to force the differential output voltage to zero. For our analyses, the input offset voltage is zero, since we have assumed everything is ideal. However, real circuits are not ideal.

Input offset voltage is primarily a consequence of device mismatches. The three main sources of mismatching are differences in the base-emitter areas between transistors, differences in base doping between the transistors and differences in the values of the collector resistances. The result of these differences is that the currents flowing in Q1 and Q2 are different, and so the V_{BE}s required for each transistor are different. We can lump the changes in current due to base doping and emitter area variations together and deal with them as a variation of saturation current I_S. We can then write

$$V_{OS} = V_T \left(-\frac{\Delta R_C}{R_C} - \frac{\Delta I_S}{I_S} \right)$$

The difference factors are random in nature and need to be dealt with in statistical fashion.

Input offset voltage will vary with temperature. This can be quantified by assuming the difference factors are independent of temperature. The change in offset voltage is then obtained by taking the derivative of V_{OS} with respect to temperature. This amounts to taking the derivative of KT/q:

$$\frac{dV_T}{dT} = \frac{d}{dT}\left(\frac{KT}{q}\right) = \frac{K}{q} = \frac{V_T}{T}$$

That is, the drift in offset voltage measured at a particular temperature will be equal to the offset voltage divided by the temperature with units of volts per degree Centigrade.

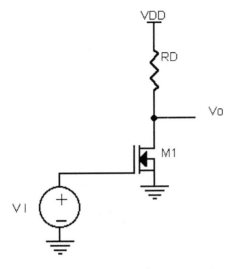

Figure 5.25   Common-source amplifier.

5.9   The MOS Case: The Common-Source Amplifier

The MOS equivalent of the common-emitter amplifier is the common-source amplifier whose schematic is shown in Figure 5.25. In this case, we again consider the amplifier with resistive loading. The drain resistor is denoted R_D. We will begin our analysis of this stage by again considering the large signal performance.

With $V_I = V_{GS} = 0$, the transistor is cut off, and no current flows. Vo is equal to V_{DD}. As V_I increases, the threshold voltage (V_{TH}) of the FET is exceeded, and the transistor operates in the saturation region and current flows in the transistor. Increasing the value of V_{GS} increases the current, and the output voltage decreases until $V_o = V_{DS} = V_{GS} - V_{TH}$. At this point, the FET enters the ohmic region. This transfer characteristic is shown in Figure 5.26.

Once the transistor is operating in the ohmic region, its output resistance decreases dramatically. This results in a decrease in the transistor gain. For this reason, we will assume our transistors operate in the saturation region.

The small-signal equivalent circuit is shown in Figure 5.27. Note that we have not included the source associated with the body diode since the source and body are both grounded.

As R_D approaches infinity, the gain of this amplifier approaches

$$A_V = -g_m r_o$$

In the ideal case, input resistance $R_I = \infty$ which implies the MOS device

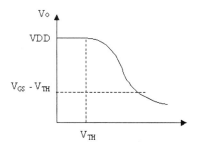

Figure 5.26   The transfer characteristic for the common-source amplifier shown in Figure 5.25.

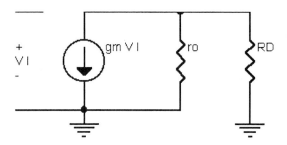

Figure 5.27   Small signal equivalent circuit for the common-source amplifier shown in Figure 5.25.

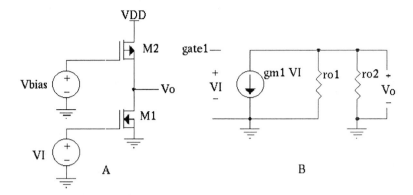

Figure 5.28   A CMOS inverter is shown in A. The small signal equivalent circuit is shown in B.

has infinite current gain.

It is worth noting that g_m for MOS transistors is dependent on the square root of drain current while the output resistance is dependent on the inverse of drain current. Thus the voltage gain will vary as $\sqrt{I_D}$. This contrasts to the bipolar case where gain is independent of collector current.

5.10   The CMOS Inverter

A special case of the common-source amplifier is the CMOS inverter. This circuit uses one transistor as the amplifier and a second transistor as the load. The second transistor is biased to provide a constant current. In this case, the amplifier voltage gain is still of the form

$$A_V = -g_m R_o$$

but $R_o = r_{o1} \parallel r_{o2}$. This is shown in Figure 5.28.

5.11   The Common-Source Amplifier with Source Degeneration

Source degeneration is typically not used in MOS amplifier design. The major drawback with this configuration is that it lowers amplifier gain. Since MOS amplifiers typically have low gain, degeneration is usually not desired. Degeneration does provide an increase in output resistance, however, and in some cases this can be a desirable feature. The schematic and small-signal equivalent circuit are shown in Figure 5.29. Note that in this case, the body effect transconductance must be consid-

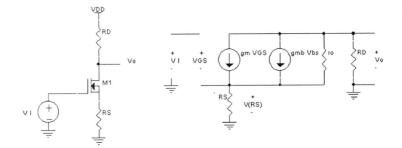

Figure 5.29 A common-source amplifier with source degeneration.

ered, and indeed has an effect on the output resistance. As in all MOS
transistors, we assume that input resistance is infinite.

If we assume that R_D again approaches infinity, the transconductance
can be calculated:

$$V_I = V_{GS} + V(R_S)$$

$$V_{bs} = -V(R_S)$$

$$V(R_S) = I_D R_S$$

$$I_D = g_m V_{GS} - g_m V_{R_S} = g_m(V_I - V(R_S)) - g_{mb} V(R_S)$$

so

$$I_D + I_D R_S(g_m + g_{mb}) = g_m V_I$$

and

$$G_m = \frac{I_D}{V_I} = \frac{g_m}{1 + (g_m + g_{mb})R_S}$$

Using a test current tied to the drain, we short the gate connection to
ground and solve for the output resistance. We obtain the following
equations:

$$V_X = I_X R_S + r_o(I_X - g_m V_{GS} - g_{mb} V_{bs})$$

$$V_{GS} = -V(R_S)$$

$$V_{bs} = -V(R_S)$$

$$V(R_S) = I_X R_S$$

and

$$R_o = \frac{V_X}{I_X} = r_o(1 + (g_m + g_{mb})R_S) + R_S$$

It is important to note that as R_S increases, R_o continues to increase.
In the bipolar case, as R_E increases, Ro approaches an upper bound of
βr_o.

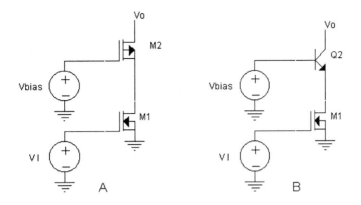

Figure 5.30   An MOS cascode amplifier is shown in A and a biCMOS cascode amplifier is shown in B.

5.12   The MOS Cascode Amplifier

Cascoding is a widely used technique in MOS technology. It is used in amplifiers and current sources to increase the output resistance. The MOS cascode amplifier is shown in Figure 5.30A. For this amplifier, $G_m = g_{m1}$, input resistance is infinite and the output resistance is obtained in the same manner as for the common-source amplifier with source degeneration:

$$R_o = r_{o2}(1 + (g_{m2} + g_{mb2})r_{o1}) + r_{o1}$$

A biCMOS alternative to the MOS cascode is shown in Figure 5.30B. This circuit has infinite input impedance and the transconductance of the bipolar device is much higher than that of a MOS device. This results in better high-frequency performance.

5.13   The Common-Drain (Source Follower) Amplifier

The source follower is shown schematically in Figure 5.31A. The small signal equivalent circuit is provided in Figure 5.31B. We again assume that r_o is very large and can be neglected.

We first note that $V_{GS} = V_I - V_o$ and that $V_{bs} = -V_o$. Applying KCL at the V_o node yields

$$g_m V_{GS} + g_{mb} V_{bs} - \frac{V_o}{R_L} = 0$$

or

$$g_m(V_I - V_o) - g_{mb} V_o - \frac{V_o}{R_L} = 0$$

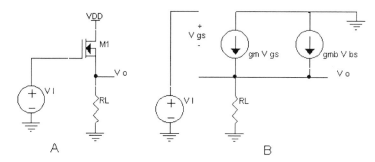

Figure 5.31   Common-drain amplifier and small signal equivalent circuit.

Rearranging, we obtain

$$A_V = \frac{V_o}{V_I} = \frac{g_m}{g_m + g_{mb} + \frac{1}{R_L}}$$

Maximum gain is realized as load resistance becomes very large, but the body effect limits this amplifier to voltage gain significantly less than unity. Values of 0.8 to 0.9 are common. This may be improved if a well process is available. In this case, the follower transistor can be placed in a dedicated well and the follower source tied to the well. In this case, the body effect transconductance is inactive and the gain reduces to

$$A_V = \frac{V_o}{V_I} = \frac{g_m}{g_m + \frac{1}{R_L}}$$

which does approach unity as $R_L \to \infty$. If we now set $V_I = 0$, we can solve for the short circuit output resistance.

$$I_o = g_m V_{GS} = g_{mb} V_{bs}$$

but $V_{GS} = V_{bs} = -V_o$, so

$$I_o = -V_o(g_m + g_{mb})$$

and

$$R_o = \frac{V_o}{I_o} = \frac{1}{g_m + g_{mb}}$$

Source followers are used in the same fashion as emitter followers. They typically provide impedance transformation and level shifting.

5.14   Source-Coupled Pairs

The importance of the source-coupled pair is obvious if we consider one of the main assumptions made in analyzing ideal operational amplifier

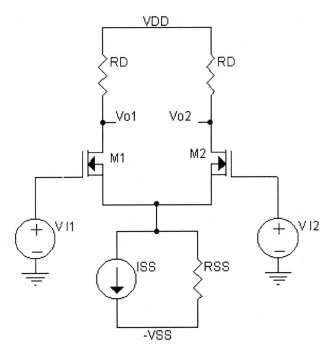

Figure 5.32   Source-coupled pair.

circuits. The assumption that no current flows into or out of the input nodes presents the circuit designer with the requirement for infinite input resistance.

The schematic for a resistively loaded n-channel FET pair is shown in Figure 5.32. We will again begin our analysis with the large signal case. We again assume both transistors are operating in the saturation region where transconductance is as high as possible, and that both transistors are identical, i.e., the same width, length and threshold voltage.

From the drain current equation for saturation we obtain

$$V_{GS1} = V_{TH} + \sqrt{\frac{I_{D1}}{K}}$$

and

$$V_{GS2} = V_{TH} + \sqrt{\frac{I_{D2}}{K}}$$

where

$$K = \frac{W}{L} \frac{\mu C_{OX}}{2}$$

We define the difference input voltage

$$\Delta V_I = V_{I1} - V_{I2} = V_{GS1} - V_{GS2}$$

We observe that the average drain current is

$$I_D = \frac{I_{D1} - I_{D2}}{2} = \frac{I_{SS}}{2}$$

and the difference current due to changes in V_{GS} is

$$\Delta I_D = I_{D1} - I_{D2}$$

This leads to

$$I_{D1} = I_D + \frac{\Delta I_D}{2}$$

and

$$I_{D2} = I_D - \frac{\Delta I_D}{2}$$

We can rearrange the equations above to determine how drain current changes for a change in the input voltage:

$$\Delta V_I = V_{GS1} - V_{GS2} = \sqrt{\frac{I_D + \frac{\Delta I_D}{2}}{K}}\sqrt{\frac{I_D - \frac{\Delta I_D}{2}}{K}}$$

Squaring both sides, and recognizing that $2I_D = I_{SS}$, we find

$$(\Delta V_I)^2 = \frac{I_{SS} - \sqrt{I_{SS}^2 - \Delta I_D^2}}{K}$$

Solving this equation for ΔID gives

$$\Delta I_D = \Delta V_I K \sqrt{\frac{2 I_{SS}}{K} - \Delta V_I^2}$$

This holds true for the case when both transistors are in saturation, or when $\Delta V_I^2 \leq I_{SS}/K$. A plot of the transfer characteristic ΔI_D vs ΔV_I is provided in Figure 5.33. Note the slope of the transfer characteristic increases as I_{SS} increases or as K decreases.

Next, we consider small-signal performance. The input resistance is infinite. The amplifier transconductance is found by taking the derivative of the change in output current with respect to the change in input voltage when evaluated at zero. That is

$$G_m = \left.\frac{d(\Delta I_D)}{d(\Delta V_I)}\right|_{\Delta V_I = 0} = \left.\frac{2 I_{SS} - 2k \Delta V_I^2}{\sqrt{2\frac{I_{SS}}{K} - \Delta V_I^2}}\right|_{\Delta V_I = 0} = \sqrt{2 K I_{SS}}$$

$$G_m = \sqrt{2\frac{W}{L}\frac{\mu C_{OX} I_{SS}}{2}} = \sqrt{2\frac{W}{L}\mu C_{OX} I_D}$$

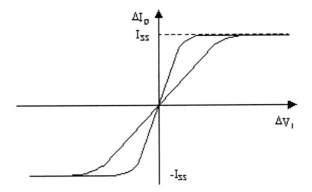

Figure 5.33   Difference in source-coupled drain currents as a function of the differential input voltage.

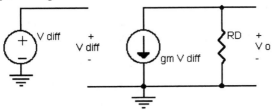

Figure 5.34   Source-coupled pair small signal equivalent half-circuit.

Thus G_m of the amplifier is equal to g_m of either transistor.

The small-signal equivalent half-circuit for the differential mode is shown in Figure 5.34. In this case, ΔV_I is denoted V_{diff}, and V_{odiff} is the differential output voltage given as $V_{o1} - V_{o2}$. We have also assumed that the output resistance is much larger than R_D, and that the source associated with the body effect is not active. Both of these circuit elements can then be ignored. It can be readily seen that the differential output voltage is

$$V_{odiff} = V_{o1} - V_{o2} = -\frac{\Delta I_D}{2} R_D - \frac{\Delta I_D}{2} R_D$$

but

$$\Delta I_D = G_m \Delta V_I = G_m V_{diff}$$

so

$$V_{odiff} = -G_m R_D V_{diff}$$

and the differential mode voltage gain of the amplifier is

$$A_{DM} = \frac{V_{odiff}}{V_{diff}} = -G_m R_D$$

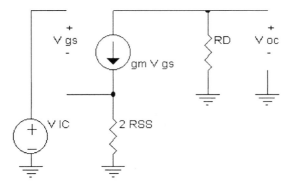

Figure 5.35   Small signal common-mode circuit for the source-coupled pair.

As in the bipolar case, the voltage gain is cut in half if the output is taken from only one side of the amplifier.

The small-signal common-mode equivalent half-circuit is shown in Figure 5.35. We have again assumed the effects of output resistance and of the body effect source are negligible. By repeating the analysis performed for the bipolar case, we find the common-mode gain for the MOS case is

$$A_{CM} = -\frac{G_m R_D}{1 + 2G_m R_{SS}}$$

Common-mode rejection ratio is again given as the differential-mode gain divided by the common-mode gain:

$$CMRR = \frac{A_{DM}}{A_{CM}} = 1 + 2G_m R_{SS}$$

Common-mode input range for the MOS case is that range of input voltage such that both sides of the source-coupled pair operate in the saturation region. Referring to Figure 5.32, we see that the lowest possible input voltage must allow for the finite voltage developed across the I_{SS} element and for the proper value of V_{GS}. V_{GS} must exceed the threshold voltage by a value ΔV given by $\Delta V = \sqrt{I_D/K}$. Since saturation requires that V_{DS} be greater than $V_{GS} - V_{TH}$, the finite voltage across the current element is also ΔV. The minimum allowable input voltage for proper operation is in this case $V_{TH} + 2\Delta V$.

The maximum voltage is limited by the voltage across R_D and bias requirements for the FET. Let us define a voltage $V_D = V_{DD} - I_{SS}R_D$. The FET needs ΔV across V_{DS} and $\Delta V + V_{TH}$ across V_{GS}. Thus, pulling the FET gate up past $V_D + V_{TH}$ will force the transistor into the triode region. This is then the maximum input voltage.

Input offset voltage in the MOS case is somewhat more complicated than for the bipolar case. In addition to the errors due to drain current mismatch and load resistor mismatch, there are threshold voltage mismatch and ΔV mismatch to consider as well. Without going through all the mathematics, MOS input offset voltage is found to be approximately

$$V_{OS} = \Delta V_{TH} + \frac{\Delta V}{2} \left(-\frac{\Delta R_D}{R_D(AVE)} - \frac{\frac{\Delta W}{L}}{\frac{W}{L}_{AVE}} \right)$$

Note that errors in load resistance or in W/L are scaled by the ΔV term. The scaling factor in the bipolar case is the thermal voltage V_T, which is usually smaller than $\Delta V/2$ by a factor of five or ten. An error term found in the MOS case that is not present in the bipolar case is the error in the threshold voltage. This offset is independent of bias current and has a strong dependence on wafer processing. Low contaminant particle count in the fabrication area and uniformity of silicon-silicon dioxide interfaces on the wafers will help to minimize this error term.

It should be noted that MOS devices will usually have worse offset performance due to lower transconductance to bias current ratios. $V_T = I_C/g_m = 26\ mV$ in bipolar, while $\Delta V/2 = I_D/g_m$ in MOS can be several hundred millivolts.

5.15   Chapter Exercises

1. For the common-emitter amplifier in Figure 5.1, let $R_C = 10\ K\Omega$, $V_A = 125V$, $\beta = 200$ and $I_C = 200\mu A$. Find input resistance, transconductance, output resistance and voltage gain.

2. For the circuit in problem 1, add a load resistance of 100 $K\Omega$ to ground. What effect does the load resistance have on input resistance, output resistance and voltage gain? Comment on the output signal (maximum value, minimum value, distortion of the waveform).

3. For the circuit in problem 1, add a source resistance $R_S = 50\Omega$. What affect does this have on input resistance, output resistance and voltage gain?

4. Use the common-base amplifier from Figure 5.9 with $R_C = 100\ K\Omega$, $I_C = 100\mu A$ and $\beta = 200$. Calculate input resistance, output resistance and transconductance.

5. Use the emitter-follower circuit from Figure 5.10 with $R_S = 10\ K\Omega$, $R_L = 10\ K\Omega$, $\beta = 200$ and $I_C = 100\mu A$. Find the input resistance (not including R_S), output resistance (include the effects of R_L) and voltage gain.

6. For the Darlington pair shown in Figure 5.14, replace the current source I_{bias} with a 10 $K\Omega$ resistor. Assume $\beta = 200$ and $I_o = 2mA$. What currents flow in each transistor? Find the input resistance, output resistance and transconductance.

7. Use the cascode amplifier shown in Figure 5.15. Assume $I_o = 100uA$. Find the input resistance, output resistance and transconductance. What is the minimum voltage on the output that maintains both transistors operating in the linear region?

8. Use the emitter-coupled pair circuit in Figure 5.18. $V_{CC} = -V_{EE} = 5V$, $I_{EE} = 10\mu A$, $R_{EE} = 1M\Omega$ and $R_C = 20$ $K\Omega$. Find A_{DM}, A_{CM}, $CMRR$, input resistance, output resistance and common-mode input range.

9. Show that taking a single-ended output from the emitter-coupled pair (i.e., using V_{o1} or V_{o2} alone) reduces the voltage gain by half.

10. For the source-coupled pair, $I_{SS} = 50$ μA, $W/L = 50$, $\mu C_{ox} = 20$ $\mu A/V^2$, and $R_D = 50$ $K\Omega$. Find transconductance, A_{DM}, C_{CM}, $CMRR$ and common-mode input range. Ignore r_o in your calculations.

11. The common-source amplifier in Figure 5.25 has a load resistance $R_D = 30$ $K\Omega$. If $KP = 600\mu A/V^2$, $\lambda = 0.04$ V^{-1} and $V_{GS} - V_{TH} = 0.5\,V$, find the voltage gain and output resistance.

12. For the circuit in problem 11, add a 1 $K\Omega$ source degeneration resistance. Assume $V_{bs} = 0$. What effect does this have on voltage gain and output resistance?

13. For the CMOS inverter shown in Figure 5.28, the bias current is 60 μA. $KP = 600$ $\mu A/V^2$ for the n-channel FET and 300 $\mu A/V^2$ for the p-channel FET. $\lambda_N = 0.004$ V^{-1} and $\lambda_P = 0.06$ V^{-1}. Find voltage gain and output resistance.

14. For the source follower shown in Figure 5.31A, $R_L = 50$ $K\Omega$, $K = 600$ $\mu A/V^2$ and $\Delta V = 0.5\,V$. Find voltage gain and output resistance. Assume $V_{bs} = 0$ and ignore r_o. Next, assume $g_{mb} = 5$ $\mu A/V$. How does this affect voltage gain and output resistance?

15. For the MOS cascode amplifier shown in Figure 5.30A, $I_D = 100$ μA, $V_{GS} = 1.3V$, and $V_{TH} = 0.85V$. $K = 600$ $\mu A/V^2$ and $\lambda = 0.04V^{-1}$. Find the transconductance and output resistance.

16. For the emitter-coupled pair, replace the resistive load with an active load as shown in Figure 5.36. Use small signal analysis

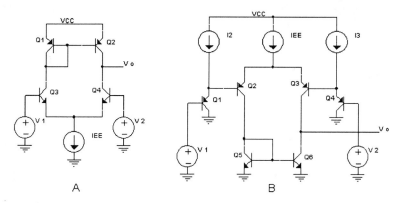

Figure 5.36   Schematics for exercises 16 and 17.

to show the voltage gain for this amplifier, $A_V = -g_m R_o$, where $R_o = r_o(p) \parallel r_o(n)$. Which is the inverting input? Determine the common-mode input range. Does the common-mode input range ($CMIR$) include ground or V_{CC}? Why or why not?

17. The PNP emitter-coupled pair shown in Figure 5.36 includes ground in the common-mode input range ($CMIR$). Why? What is the purpose of sources I_2 and I_3?

References

[1] Baker, R. Jacob, et al, *CMOS Circuit Design, Layout and Simulation*, IEEE Press, New York, c. 1998.

[2] Gray, Paul R., and Mayer, Robert G., *Analysis and Design of Analog Integrated Circuits*, 2nd edition, John Wiley and Sons, Inc., New York, c. 1984.

[3] Millman, Jacob, and Grabel, Arvin, *Microelectronics*, 2nd edition, McGraw-Hill Book Company, New York, c. 1987.

[4] Soclof, Sidney, *Analog Integrated Circuits*, Prentice-Hall, Inc., Englewood Cliffs, N.J., c. 1985.

chapter 6

Comparators

A comparator is a functional circuit block that compares the relative levels of two signals. The comparator's output signal provides a logic "1" or "0" depending on the result of the comparison.

The ideal comparator is perfectly accurate and instantly provides the correct output. Again, real life intrudes and gives us limits. Finite gain and non-zero propagation delays result in delays between the input signal being applied and the output signal being available. Input offsets result in errors in the comparison. However, these problems can be overcome, and many circuits can be built from the basic comparator. Oscillators, d/a converters and a/d converters all use some form of comparator, and operational amplifiers are comparators with frequency compensation.

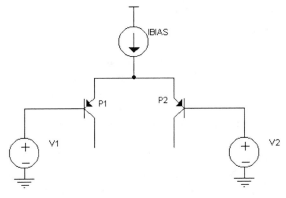

Figure 6.1   A typical comparator input stage.

Consider the emitter-coupled pair shown in Figure 6.1. This is a standard comparator input stage. We can use Kirchoff's Voltage Law on the loop starting at V_1's ground and ending at V_2's ground to obtain

$$V_1 - V_{be}(P1) + V_{be}(P2) - V_2 = 0$$

We can express the base-emitter voltages of the transistors using the diode equation:

$$V_{be}(P1) = V_T \ln\left(\frac{I_{C1}}{I_{S1}}\right)$$

$$V_{be}(P2) = V_T \ln\left(\frac{I_{C2}}{I_{S2}}\right)$$

If we assume the pnp transistors are identical, then $I_{S1} = I_{S2}$, and the ratio of collector currents is given as

$$\frac{I_{C1}}{I_{C2}} = exp\left(\frac{V_1 - V_2}{V_T}\right) = exp\left(\frac{V_{dif}}{V_T}\right)$$

This circuit acts as a linear amplifier for a small range around $V_{dif} = 0$. However, when V_{dif} exceeds several tens of millivolts, this circuit acts as a pair of complementary switches. We have previously analyzed current mirrors and found that a transistor's collector current approximately doubles for an increase of $18mV$ in the magnitude of the base-emitter voltage. In the circuit of Figure 6.1, the total current available is equal to the bias current. When $V_1 = V_2$, both transistors are conducting. If we assume β is infinite, $I_{C1} = I_{C2} = I_{BIAS}/2$. If V_1 is $18mV$ lower than V_2, the magnitude of P1's V_{be} is $18mV$ greater than that of P2, and P1 will conduct twice the current that P2 does. Thus, $I_{C1} = (2/3)I_{BIAS}$ and $I_{C2} = (1/3)I_{BIAS}$. If $V_{dif} = 100mV$, the ratio of collector currents is nearly 50.

Figure 6.2 shows how the emitter-coupled pair can drive an output stage. Our analysis assumes that the output node drives a high impedance load and that our ideal transistors are not turned on for V_{be} less than 0.7V. β is also assumed to be infinite. A V_{be} developed across R1 will turn transistor N1 on. N1 will eventually saturate and V_{out} is pulled low. The absence of a V_{be} leaves N1 off, and resistor R_{LOAD} pulls V_{out} high.

We can start our analysis by noting that $V_2 = 1V$. We arbitrarily begin by assuming that $V_1 = 0V$. In this case, the voltage at the pnp emitters is clamped to about 0.7V due to the V_{be} of P1. This means that P2 is cut off and all of I_{BIAS} flows in P1. There is no current flow in R1 and so $V_{be}(N1) = 0V$. This results in $V_{out} = VCC = 12V$.

We let V_1 begin to rise. Eventually, V_1 is about $100mV$ below V_2. At this point, P2 is conducting about $2\mu A$ while P1 conducts about $98\mu A$. The collector current of P2 results in about $20mV$ being dropped across R1. This voltage is not sufficient to turn N1 on, so V_{out} stays high. V_1

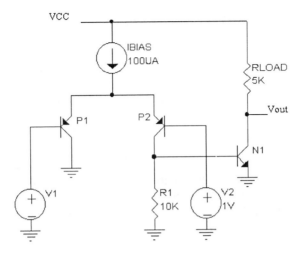

Figure 6.2 A simple comparator.

continues to rise until $V_1 = V_2$. This means that I_{BIAS} is split equally between P1 and P2, and so about $50\mu A$ flows through R1. $V_{be}(N1)$ is now about $500mV$. N1 is still cut off and V_{out} is still high. When V_1 is $18mV$ higher than V_2, P2 conducts twice the collector current of P1. This is approximately $66\mu A$. $V_{be}(N1)$ is now about $660mV$. N1 is still cut off according to the rules of our analysis, but it is on the verge of turning on. When V_1 is about $22mV$ higher than V_2, $I_{C2} = 70\mu A$ and N1 turns on, saturates and pulls V_{out} low. The transfer characteristic described above is shown in Figure 6.3A.

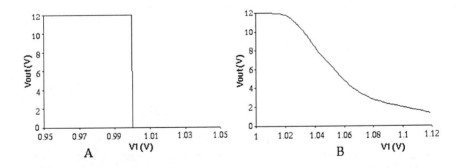

Figure 6.3 A. Ideal comparator transfer function. B. "Real Life" comparator transfer function.

Let us review some of the assumptions in our analysis. First, the fact that transistors begin to conduct before $V_{be} = 0.7V$ will result in a "soft" turn on characteristic for N1. This results in the V_{out} transition having a more gradual slope. Also, the point at which we consider V_{out} a logic "0" is important. Base current needed to drive N1 must be sufficient to cause the required voltage drop across R_{LOAD}. This means some additional offset will occur due to the current needed to drive N1. Setting $V_2 = 1V$ is another important decision. This prevents P2 from being forced into saturation when N1 is turned on. Let us consider what would happen if V_2 were set to $500mV$. When $V_1 = 500mV$, both transistors would conduct $50\mu A$ and the voltage drop across R1 would be $500mV$. In this case, $V_{BC}(P2) = 0$. When V_1 is changed to $522mV$, the voltage drop across R1 is $700mV$ and $V_{BC}(P2) = 300mV$. As V_1 increases, $V_{BC}(P2)$ is forced to decrease until P2 saturates. When this occurs, the parasitic transistors associated with P2 shunt current to ground. This may impact correct operation of the circuit. Finally, we notice there is an offset on the order of $50mV$ from the desired switching point. A corrected transfer characteristic, shown in Figure 6.3B, shows the effects of $I_S = 200\text{E-}18$ and $\beta = 100$.

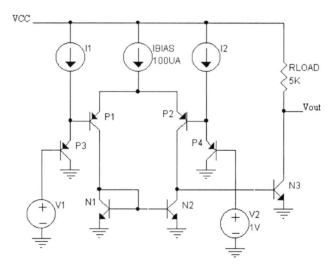

Figure 6.4 An improved comparator.

We can enhance the performance of the circuit in Figure 6.2. The circuit in Figure 6.4 has two distinct improvements. First, we have added level shifting transistors P3 and P4. These transistors add an extra V_{be} between the inputs and the emitter-coupled pair. This allows the comparator to function properly even if $V_1 = 0V$. The additional current mirrors I_1 and I_2 provide current to charge the emitter-base capacitance of the level-shifting transistors. This increases the speed at which the input stage can respond to input transients. If these currents were not provided, the base currents of P1 and P2 would have to charge the level-shifter capacitances. Since the base currents are small, the time required to respond to a fast transient would increase. Second, we have added a current mirror active load. The active load results in the reduction of the offset inherent in the Figure 6.2 circuit due to the presence of R1.

We can examine operation of this improved circuit by again assuming $V_1 = 0V$. P1 and P3 are turned on while P2 and P4 are cut off. All of I_{BIAS} flows in P1 and N1. N2 attempts to mirror I_{BIAS}, but no current flows through P2. N2 saturates and holds N3 in cutoff. No current flows in R_{LOAD} and V_{out} is held high. As V_1 approaches $1V$, the currents in P1 and P2 approach $I_{BIAS}/2$. When they are equal, N2 sinks all the current provided by P2, and N3 is on the verge of turning on. As V_1 rises, P2 conducts more than P1, and N2 does not sink all the current provided by P2. Base current is then available to drive N3. N3 begins to conduct, pulling current through R_{LOAD} and causing V_{out} to drop.

Example

Evaluate the transfer characteristic of the circuit shown in Figure 6.4. Assume transistor $\beta = 100$.

We begin our analysis by noting that $V_{out} = 12V$ for $V_1 = 0V$. Next we will find the equilibrium point for P2 and N2. At this point, their collector currents are equal and no current is available for base drive to N3. We note the following:

$$I_C(P2) = I_C(N2) = I_C(N1)$$

$$I_C(P1) = I_C(N1) + 2I_B$$

$$\beta = 100$$

$$I_C(P1) + I_C(P2) = 100\mu A$$

Rearranging, we obtain

$$1.02I_C(N1) + I_C(N1) = 100\mu A, \quad \text{or} \quad I_C(N1) = 49.505\mu A$$

Then $I_C(N2) = I_C(P2) = 49.505\mu A$ and $I_C(P1) = 50.495\mu A$. The difference between V_1 and V_2 is

$$V_{dif} = (26mV)ln\left(\frac{50.495\mu A}{49.505\mu A}\right) = 0.5mV$$

Since P1 is conducting a larger current, equilibrium occurs when $V_1 = 0.9995V$. Any voltage larger than this will result in N3 conducting and V_{out} being less than 12V.

At $V1 = 1V$, the collector currents in P1 and P2 are equal. Then

$$I_C(P1) = 50\mu A = 1.02 I_C(N1), \quad \text{or} \quad I_C(N1) = 49.02\mu A$$

Thus, $I_C(P2) = 50\mu A$ and $I_c(N2) = 49.02\mu A$. This gives $I_B(N3) = 0.98\mu A$. From this, we find $I_C(N3) = \beta I_B(N3) = 98\mu A$. Finally we obtain

$$V_{out} = 12V - 5K\Omega I_C(N3) = 11.51V$$

We can work backwards to find some more points. Consider the case when N3 is saturated. Assume $V_{out} = 0.3V$. Then

$$I_C(N3) = \frac{11.7V}{5K\Omega} = 2.34mA$$

Then

$$I_B(N3) = \frac{2.34mA}{100} = 23.4\mu A$$

We can then write

$$I_C(P2) - I_C(N2) = 23.4\mu A$$

$$I_C(P1) = 1.02 I_C(N2)$$

$$I_C(P2) = 100 - I_C(P1)$$

Rearranging, we have

$$I_C(P2) - \frac{I_C(P1)}{1.02} = 23.4\mu A$$

$$100\mu A - I_C(P1) - \frac{I_C(P1)}{1.02} = 23.4\mu A$$

$$I_C(P1) = 38.679\mu A$$

Then $I_C(P2) = 61.321\mu A$ and $V_{dif} = 0.026V ln(61.321/38.679) = 11.98mV$. The same method can be used to identify more points in the transfer curve (plotted in Figure 6.5).

At this point, our comparator function is fairly well defined. However, we still have some limitations. For instance, the present design still switches slowly for small values of V_{dif}. Also, random noise on either the input or the reference can result in incorrect output switching. This can be corrected by providing some hysteresis to the reference.

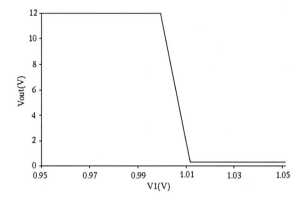

Figure 6.5   Comparator transfer function plotted using text example hand calculation data.

6.1   Hysteresis

Hysteresis involves adding positive feedback circuitry to modify the threshold reference voltage. The threshold is modified in the opposite polarity from the direction of approach of the input signal. That is, if V_1 passes through the threshold set by V_2 while increasing, the value of V_2 will decrease. Similarly, if V_1 passes through the threshold set by V_2 while decreasing, the value of V_2 will increase. The amount of deviation from the unmodified reference is called hysteresis voltage. Hysteresis of several hundred millivolts helps to remove switching due to noise and ensures fast, clean output transitions. The circuit shown in Figure 6.6 adds hysteresis to the basic comparator function. Our $1V$ reference is now generated using a current source and resistor R1. The hysteresis network consists of N4 and R2.

6.1.1   Hysteresis with a Resistor Divider

Let us begin our circuit analysis by assuming $VCC = 12V$ and $V_1 = 0V$. Then I_{BIAS} sinks through P1 and N1. P2 is cut off while N2 is saturated, resulting in both N3 and N4 being cut off. If N4 is cut off, there is no current flowing through R2, and the reference voltage V_2 is the product of I_{REF} and R1. As V_1 rises, it reaches the threshold level. As current begins to flow to the bases of N3 and N4, N4 begins to turn on. This causes current to be diverted away from R1, causing V_2 to drop. Once this occurs, the magnitudes of the V_{be}s for P2 and P4 increase, resulting in more base current provided to N4, in turn diverting more current from R1. This cycle continues until N4 is saturated. The value of R2 was chosen such that $V_2 \approx 500mV$ when this occurs. The total time it takes for this to occur is very small, possibly in the tens of nanoseconds.

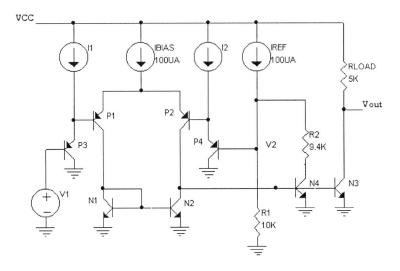

Figure 6.6   Comparator with resistor divider hysteresis.

As V_2 decreases, V_{out} also decreases, since N3 and N4 are driven in parallel. However, the transition of V_{out} is now very sharp and well defined. Additionally, V_1 must now decrease to the $500mV$ level in order for the output to transition high again. This effectively guards against false switching due to noise. The transfer function for this circuit is shown in Figure 6.7.

6.1.2   Hysteresis from Transistor Current Density

There are several ways in which hysteresis can be added to a comparator. Using a change in the reference voltage is one method. Another alternative is to use a change in transistor collector current density. The circuit in Figure 6.8 illustrates this possibility. Resistor R1 and the current mirror made up of N3 and N4 set the comparator "long-tail" current at about $100\mu A$. If we begin our analysis by assuming that $V(IN^+)$ is much higher than $V(IN^-)$, we can assume that N2 sinks all of the $100\mu A$ required by N4. P2A is configured as a mirror. It carries $100\mu A$, with the result that P2B, P2C and P2D also try to conduct $100\mu A$. P2D provides $100\mu A$ to a load. We assume the load is high impedance such that V_{out} is pulled up until P2D saturates. However, N1 is assumed to be cut off, so the collector currents of P2B and P2C have nowhere to go. These transistors saturate, cutting P1A off. P1B, P1C, P1D, N5 and N6 are all cut off as a result, and V_{out} should indeed be pulled high.

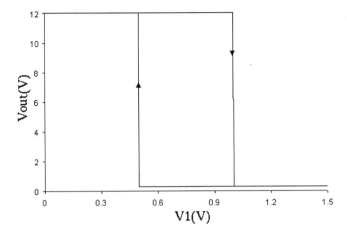

Figure 6.7 Transfer characteristic of comparator with resistor divider hysteresis.

We now let $V(IN^+)$ decrease. At some point, $V(IN^+) = V(IN^-)$. When this occurs, both N1 and N2 conduct $50\mu A$. This results in P2B and P2C trying to mirror a total of $100\mu A$. Since N1 is only sinking $50\mu A$, the P1 mirror is still cut off and V_{out} is still high. Note, however, that the drive capability of the output has been decreased from $100\mu A$ to $50\mu A$.

When $V(IN^-)$ is $18mV$ higher than $V(IN^+)$, N1 sinks $66.66\mu A$ while N2 sinks $33.33\mu A$. At this point, N1 is capable of sinking all the current provided from P2B and P2C. The comparator output is still capable of $33.33\mu A$ of pull-up current.

If $V(IN^-)$ increases further, the current in N2 will decrease, and N1 will begin to pull current from P1A. P1B and P1C will begin to source current to N2's collector, reducing the current in P2A. This results in N1 pulling more current from P1A, and the circuit quickly transitions from sourcing current to sinking current. P1D drives the mirror made up of N5 and N6, pulling the output node down.

The same analysis applies in the reverse case. The low-to-high output transition will occur when $V(IN^+)$ is $18mV$ higher than $V(IN^-)$. The total hysteresis is then $36mV$, $V_{REF} \pm 18mV$. The transfer function is shown in Figure 6.9. A note of caution is useful here. In designing comparators with hysteresis, it is important to ensure that the hysteresis circuit has changed state before the output is allowed to change state. This guarantees that a clean output transition occurs. If the output changes state before the hysteresis, it is possible to get output "chatter" as the comparator changes state. Chatter refers to several rapid high-to-low-to-high output changes.

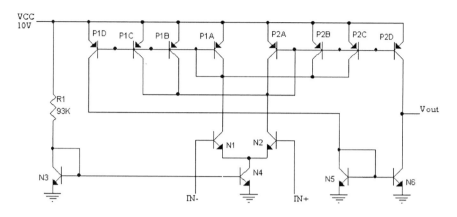

Figure 6.8   Comparator with current density hysteresis.

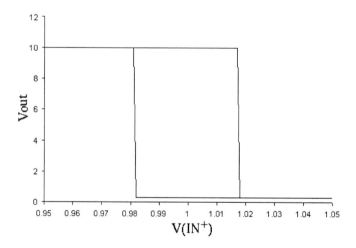

Figure 6.9   Transfer characteristic of the comparator with current density hysteresis.

6.1.3   Comparator with V_{be}-Dependent Hysteresis

Another alternative for providing hysteresis is the circuit shown in Figure 6.10. This circuit uses transistor V_{be} to provide the change in the comparator reference. Let us again start our analysis by considering the

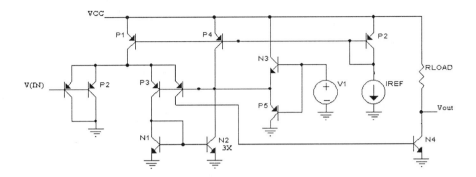

Figure 6.10   Comparator with V_{be}-dependent hysteresis.

case when $V(IN)$ is low. In that case, P2 conducts a current equal to I_{REF}. P3 is cut off, as are N1, N2 and N4. V_{out} is pulled high through R_{LOAD}. P4's collector current pulls the base of P3 up until P5 turns on. This occurs when the voltage at P3's base is equal to $V_{REF} + V_{be}$. This is the upper reference level that $V(IN)$ must cross in order for the output to change states. As $V(IN)$ crosses this level, P3 begins to conduct. P3 has two collectors, so the collector current splits. Half goes to drive the output transistor N4. The other half is gained up by a factor of 3 in the mirror made up of N1 and N2. This pulls P4's collector current to ground. However, N3 is sized to sink an additional $I_{REF}/2$. This current is provided when P3's base node voltage falls to $V_{REF} - V_{be}$. At that point, N3 clamps the lower reference level. The transfer characteristic looks like the one shown in Figure 6.9, but it has a range of $\pm V_{be}$ centered on V_{REF}.

6.2   The Bandgap Reference Comparator

We have analyzed the operation of the bandgap reference in previous chapters, but it is possible to use a crude bandgap as a comparator. Such a circuit is shown in Figure 6.11. We see the familiar ΔV_{be} npn pair, a pnp mirror, current setting resistor R2 and gain resistor R3. We have simply added an output stage consisting of P3, N3, and two resistors. R1 serves to limit the current driving the bases of N1 and N2. This helps to keep them out of hard saturation, maintaining correct operation of the circuit and limiting the input bias current. Note that N1 must be a multiple transistor. If transistor N2 has an emitter area of "X", then N1 must have an emitter area of "KX". Values of 2, 4, and 8 are good choices for "K", since they lend themselves to layouts that

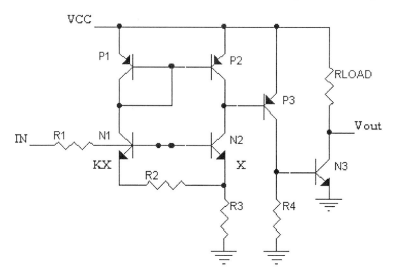

Figure 6.11   Bandgap reference comparator.

ensure good matching.

With $V(IN) = 0V$, transistors N1 and N2 are cut off. P1 and P2 are off, and R4 holds N3 in cut off. As $V(IN)$ begins to rise, N1 turns on faster than N2. P1 begins to source current, which P2 tries to mirror. N2 can't sink all the current, so P2 saturates and holds P3 in cutoff. As $V(IN)$ reaches the bandgap threshold, N2 catches up with N1. It begins to sink more current than P1 provides. This pulls current out of the base of P3, forcing it to drive R4 and N3, and the output is pulled low. The turn-on threshold is approximately given by

$$V_{on} \approx \frac{R3}{R2} 2V_T ln K + V_{be}$$

This type of comparator does not provide hysteresis, but the accuracy of the threshold is fairly good. The circuit is a good choice when a particular value of threshold is needed and hysteresis isn't a concern.

6.3   Operational Amplifiers

An operational amplifier has been described as an emitter-coupled pair input stage followed by several stages of gain and buffering. We have just seen that a comparator has essentially the same design. The two functional blocks differ in how they are used and in the design constraints based on those differences. A comparator can be viewed as an operational amplifier run in an open-loop configuration. Equivalently, an op-amp can be viewed as a comparator run in a closed-loop configuration.

The main difference between an op-amp and a comparator is that the op-amp must be designed so it provides stable operation over frequency. This requires some sort of frequency compensation circuitry. A single Miller-effect capacitor is usually sufficient for this purpose, but more complex circuitry may be required as the amplifier gain-bandwidth product requirements increase. Many texts exist which provide insight into the tradeoffs of amplifier frequency compensation, and we will not attempt to duplicate their work. Circuits shown in the following pages are examples that work in silicon with the processes used by the authors. These circuits were chosen to highlight applications where comparators and op-amps can be used.

6.4 A Programmable Current Reference

The temperature compensated voltage reference is one of the fundamental building blocks of microelectronics. However, the design of temperature stable current sources can be a tedious exercise. Also, once designed, the temperature-stable current source is usually limited to a single current value.

It is often desirable for a current reference to be adjusted over a wide range of current while still maintaining temperature-invariant performance. The circuit in Figure 6.12 meets this requirement. This circuit is based on the premise that a temperature-invariant current can be obtained by dropping a temperature-compensated voltage across a temperature-invariant resistor. Resistors with very low temperature coefficients are common and inexpensive, and bandgap references are easy to build, so this is a practical approach.

The circuit in Figure 6.12 has several sub-blocks. The bias circuitry consists of P1 through P4, N1 and R1. A temperature compensated reference voltage, shown as $V_{REF} = 2V$, is applied to the base of N1. A small operational amplifier, often referred to as an error amplifier, is made up of P5, P6, N2 through N5, and C1. Finally, P7 through P9 make up the output current reference.

The base of P6 serves as the inverting input to the error amp. The error amp tries to force $2V$ on the inverting input in order for it to be balanced. If P6's base is less than $2V$, P6 drives the npn active load and N2 diverts current from N4's base. N4's collector current decreases, and P4 pulls up on the base of N5. This causes N5's emitter voltage to rise, which puts the error amplifier back into balance. If P6's base voltage is greater than $2V$, the drive to the active load decreases and more current flows to the base of N4. N4 then pulls the base of N5 down, reducing

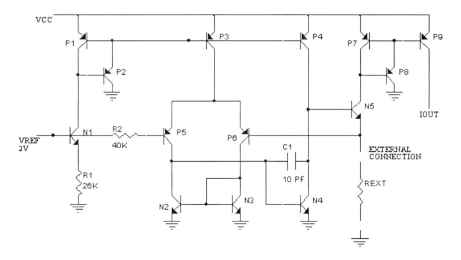

Figure 6.12   An externally-adjustable current reference.

N5's emitter voltage and again restoring the error amp to equilibrium. The emitter of N5 is connected to an external, zero TC resistor. This causes a temperature-stable current to flow in N5 and P7. The current is then mirrored to other on-chip circuitry as needed, as shown by P9. This circuit could be used in an oscillator, where the oscillator frequency could be set by an external resistor and capacitor. The resistor sets the current, while the capacitor serves as an integrator (as shown in the next section).

6.5 A Triangle-Wave Oscillator

Another very useful circuit block is a triangle-wave oscillator shown in Figure 6.13. The triangle wave is generated by alternately charging and discharging a capacitor with a constant current. The heart of this block is a comparator with hysteresis. We have chosen to implement the hysteresis using the resistor divider technique. P1, P2, and I_1 set the bias currents for this circuit. (As noted in the previous section, these components could be replaced with an externally-programmable current reference.) C1 serves as the integrating capacitor. The comparator is made up of P8 through P11, N1 through N3, and R2. The hysteresis network is made up of R1, R3, and N7. P7 and R4 set the comparator reference.

This circuit works by switching P3's collector current. This current charges C1 during the rising portion of the oscillation, and is shunted to

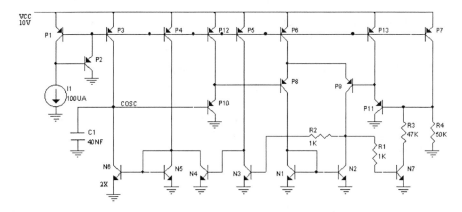

Figure 6.13   A triangle wave oscillator using a comparator with hysteresis.

ground through N6 during the falling edge of the oscillation.

At start-up, $V(COSC) = 0V$. P8 is on and P9 is cut off. The npn active load is turned on and N2 is saturated. N7 is cut off. P7 provides $100\mu A$ to R4, establishing a 5V reference. At the same time, N3 is cut off, N4 is saturated and the mirror N5/N6 is cut off. Thus, all of P3's collector current flows into C1 and $V(COSC)$ begins to increase. When $V(COSC) = 5V$, the comparator switches. P9 sources current to N7, turning on the hysteresis circuit. Putting R3 in parallel with R4 results in the threshold voltage changing to $2.5V$. N3 turns on, cutting off N4 and allowing the mirror N5/N6 to become active. Note that this mirror has a gain of two. When this mirror turns on, P3's collector current flows through N6, along with an equal amount of current discharged from C1. The integrating capacitor is thus charged with symmetric source and sink currents, resulting in a triangle waveform. The time required to charge or discharge between the two threshold levels is given as

$$\Delta T = \frac{C\Delta V}{I}$$

Since two time intervals are required to complete the oscillator period, we have

$$T = \frac{2C\Delta V}{I}$$

The oscillator frequency is $1/T$. The component values in our schematic give $T = 2ms$, or a frequency of $500Hz$.

A few comments are in order at this point. First, resistors R1 and R2 are present to reduce the possibility of a phenomenon known as current-hogging. When two saturating transistors are driven from the same source, differences in collector loading can result in one transistor being

saturated while the other is barely turned on. These resistors serve to
decouple the saturating transistors from the source. Second, note the
use of the level-shifting transistors P10 and P11. Without these transistors, it is possible that the oscillator can become "stuck" in the low state
at start-up. Finally, the triangle waveform can be changed into a sawtooth waveform by changing the symmetry of the source current to sink
current ratio. For example, changing N6 from 2X to 11X would result
in discharging C1 ten times faster. In this case, the period would con-
sist of the 1 ms charge time and the $0.1ms$ discharge time, or $T = 1.1ms$.

6.6   A Four-Bit Current Summing DAC

The circuit shown in Figure 6.14 uses both comparators and an op-amp
to implement a four bit current summing digital-to-analog converter
(DAC). N9 through N13 make up a binary-weighted current mirror.

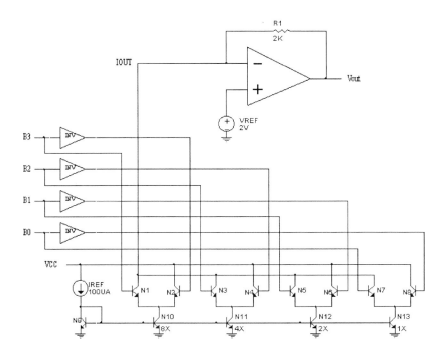

Figure 6.14   Four-bit current summing DAC.

Each emitter-coupled pair connected to the current mirror acts as a current steering comparator. If a bit is "high," the current to the mirror

is taken from the node marked "$IOUT$." If a bit is "low," the current is provided from VCC. (Note that the blocks marked INV indicate an inversion function and are not meant to indicate the presence of CMOS inverters. The choice of appropriate input circuitry is made based on input voltage levels, current levels and other design constraints.) Currents from each comparator are summed on the $IOUT$ line and used as input to the op-amp. Note that this op-amp must be capable of providing current for the entire binary mirror, and the input bias current should be low enough not to introduce significant error in the current summation. With this said, the op-amp will cause the system output to be

$$V_{out} = 2 + (2K\Omega)(IOUT)$$

where $IOUT$ is the sum of all currents flowing in the $IOUT$ line.

6.7   The MOS Case

All of the examples shown so far in this chapter have been designed in bipolar technology. The same circuits can be designed using MOS transistors, but the transfer characteristics will be slightly different. This is due to the difference in the equations that govern current flow in the transistors. The bipolar transistor's collector current is dependent on V_{be} in an exponential manner, but the drain current in a saturated MOS transistor is dependent on V_{gs}^2. In general, a pencil and paper analysis of MOS comparators is more complicated, but not impossible. We recommend using a circuit simulator in this case.

6.8   Chapter Exercises

1. Draw the transfer function for the comparator in Figure 6.2 if $\beta = 100$. Assume N1 does not turn on until $V_{be} = 0.7V$ and that $V_{be}(N1)$ does not vary from $0.7V$ when collector current changes.

2. Repeat exercise 1 but include the effects of $I_S = 200E\text{-}18A$ for N1. Comment on how the choice of the voltage level that corresponds to a logic "0" affects the threshold deviation from the reference voltage. What components could be modified to improve the accuracy of the comparator switch point? List all such components and explain the pros and cons of changing each.

3. Evaluate the transfer characteristic of the comparator in Figure 6.4 if $R_{LOAD} = 10K\Omega$. Repeat for $R_{LOAD} = 50K\Omega$.

4. Modify the comparator in Figure 6.8 to achieve hysteresis greater than $100mV$.

5. For the comparator in Figure 6.8, what would happen if two additional transistors were placed in parallel with P2B and P2C?

6. For the comparator in Figure 6.10, assume $I_{REF} = 100\mu A$, $V_1 = 2V$, $I_S = 200\text{E-18A}$ and $\Delta V_{be}/\Delta T = -2mV/°C$. What are the maximum and minimum threshold levels over the operating temperature range of -40°C to +85°C?

7. For the comparator in Figure 6.11, assume $I_S = 200\text{E-18A}$ and $\beta = 100$. Choose K, R1, R2 and R3 such that the nominal threshold is $1.250V \pm 1\%$. Size R1 to provide sufficient base current to both npns at the threshold voltage. How much input bias current flows if V(IN) increases to 5V?

8. What changes are necessary to make the oscillator in Figure 6.13 operate at 5 KHz? 50 KHz? 250 KHz? As frequency increases, what effect will propagation delays have on the circuit performance?

9. Redesign the oscillator in Figure 6.13 to produce a 50 KHz sawtooth wave that charges for 90% of the oscillator period.

10. Describe layout effects that could limit the accuracy of the DAC in Figure 6.14.

11. Design a simple circuit for the INV block in the DAC of Figure 6.14. Assume the bit signals are provided by an emitter follower output capable of pulling up to $4.3V$.

References

[1] Baker, R. Jacob, et al., *CMOS Circuit Design, Layout and Simulation*, IEEE Press, New York, c. 1998.

[2] DiTommasso, Vincenzo, *ELE536 Class Notes: Comparators and Op Amps*, Cherry Semiconductor Corporation Training Memorandum, 1997.

[3] Gray, Paul R., and Mayer, Robert G., *Analysis and Design of Analog Integrated Circuits*, 2nd edition, John Wiley and Sons, Inc., New York, c. 1984.

[4] Millman, Jacob, and Grabel, Arvin, *Microelectronics*, 2nd edition, McGraw-Hill Book Company, New York, c. 1987.

chapter 7

Amplifier Output Stages

We have seen how loading an amplifier output can affect the voltage gain and output resistance. These side effects are not desirable. Ideally, we want our amplifier to have infinite gain. While this is not possible in the real world, it is possible to minimize the effects of loading by providing one or more buffer stages between the amplifier input and the output. These buffers provide some decoupling of the amplifier gain from the output loading by impedance transformation and by providing some additional gain.

There are other considerations beyond minimizing gain sensitivity to loading. We want our amplifier output to faithfully reproduce the input signal provided to it. This means we want as little signal distortion as possible. We also want our output stage to be capable of high-frequency operation, and we want all of this accomplished with the minimum possible quiescent power consumption and in the smallest possible silicon die area.

Bipolar transistors are best suited to the tasks required of the output stage. They are fast, capable of providing high gain and Cypically introduce less noise into the signal than is the case for MOS transistors. However, bipolar transistors are often not available for use in CMOS technology, and MOS transistors can be successfully used in output stages as well.

There are many types of output stages. Typically, output stages are grouped by how the output is biased in the quiescent case. This is usually a good measure of how much power the output stage will consume and how efficient it will be in transferring power to a load. Class A output stages are biased so that the output devices are always conducting a substantial current. They dissipate large amounts of power, and ideal efficiency is limited to 25%. Class B outputs are biased to conduct no quiescent current if the input signal is zero. Also known as "push-pull" outputs, these circuits are characterized by having two active devices in the output. Each device will conduct during a half-cycle, and both active devices cannot be conducting at the same time. Class B outputs are 78.6% efficient in the ideal case, but they do exhibit signal distortion when the input signal is in the region around zero. The Class AB stage combines the best attributes of both Class A and Class B stages. The Class AB stage is biased to conduct a small quiescent current that helps to minimize the region of crossover distortion.

Output stages can be modeled as a low impedance connection between the load and the supply during conduction. Fault conditions can exist that would result in very large currents flowing in the output devices. It is possible that these conditions can result in damage to or destruction of the output devices. One way to prevent this is to include current limiting in the output stage. This circuitry allows the output stage to provide current up to some maximum value, at which point the drive to the output devices is clamped and further increase in current is not possible. We will discuss examples of Class A, Class B and Class AB output stages and will highlight some methods of overcurrent protection as well.

7.1   The Emitter Follower: a Class A Output Stage

An emitter follower can be used as an output stage as shown in Figure 7.1. Note that the circuit shown has bipolar power supplies of $\pm VCC$. This simplifies our analysis. In practice, either supply can take on any value. Our output stage is biased by transistor Q2 and resistor R_E, which can be assumed to be a current mirror conducting quiescent current I_Q.

Output stages operate with large changes in the voltage and current bias points. With this in mind, we will start our analyses with a large signal analysis, again with the assumption that all our transistors operate in the forward active region. We first observe that

$$V_I = V_{BE} = V_o$$

Since the current through Q1 can vary greatly as a result of changes in the load current, we define V_{BE} in terms of I_C:

$$V_{BE} = \frac{KT}{q} ln \left[\frac{I_C}{I_S} \right]$$

Further, we can ignore base current if β is large and approximate I_{C1} as

$$I_{C1} = I_Q + \frac{V_o}{R_L}$$

Finally, if we assume that R_L is much smaller than the output resistances of our transistors, we have the following relation between V_I and

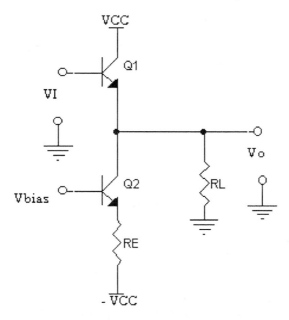

Figure 7.1   Emitter follower output stage.

V_o:

$$V_I = \frac{KT}{q} ln \left[\frac{I_Q + \frac{V_o}{R_L}}{I_S} \right] + V_o \qquad (7.1)$$

We can plot the transfer characteristics of the emitter follower output stage from this equation as shown in Figure 7.2.

Let us first examine the case where R_L is a large value. In this case, the load current will not change much. This implies that V_{BE} of Q1 doesn't change much and we can assume the first term in Equation 7.1 is constant.

The maximum output voltage is limited by two factors: how high can V_I get and either V_{BE} or VCE_{sat} of Q1. If we allow V_I to exceed VCC, the output voltage can reach $VCC - VCE_{sat}(Q1)$. However, this is usually not the case in integrated circuit design. V_I is usually derived on the IC, and so the upper limit of V_I is usually VCC.

If V_{BE} is constant, decreasing V_I results in V_o decreasing. The graph has a slope of 1 in this region. The graph crosses the x-axis at $V_I = V_{BE1}$.

As V_I decreases, V_{BE1} must remain constant in order for Q1 to remain on. The assumption that R_L is large and load current is small requires Q1 to remain on so it can provide the difference between I_Q and the load current. At some point, the current mirror Q2 will begin to saturate. Thus, the minimum value of V_o is approximately $-VCC+VCE_{sat}(Q2)$.

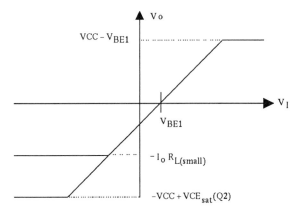

Figure 7.2   Emitter follower output stage transfer function.

The voltage across R_E is usually small enough to ignore, or it can be lumped in with $VCE_{sat}(Q2)$.

In the event that RL is small enough that load current can equal the quiescent current, the minimum output voltage is limited by the maximum possible voltage drop across the load resistor. At this point, I_Q flows entirely from the load. Q1 turns off and $V_o = -I_Q R_L$. This case is shown in Figure 7.2 as $-I_Q R_L$ (small). It is desirable that $I_Q R_L >$ VCC to ensure that the output can reproduce the input signal without distortion.

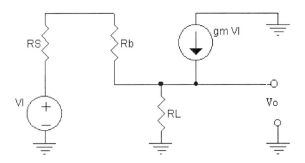

Figure 7.3   Emitter follower small signal equivalent circuit.

The low-frequency small signal equivalent circuit for the emitter follower output stage is shown in Figure 7.3. If we assume $\beta \gg 1$, voltage gain and output resistance of this circuit are given as

$$A_v = \frac{R_L}{R_L + \frac{1}{g_m} + \frac{R_S}{\beta}}$$

$$R_o = \frac{1}{g_m} + \frac{R_s}{\beta}$$

Recalling that $g_m = I_c/V_T$ and $R_o = 1/gm + R_s/\beta$, we note that both of these quantities are dependent on the collector current of Q1. We have already noted that wide swings in collector current can be expected in use as an output stage, and so the use of small signal analysis for the emitter follower output stage must be carefully considered. Reasonable estimates of both voltage gain and output resistance can be obtained by using quiescent bias current in the equations as long as the input voltage changes are moderate.

7.2   The Common-Emitter Class A Output Stage

The common-emitter circuit is another frequently used output stage. It has advantages over the emitter follower stage in that voltage gain is possible and the output voltage can swing closer to the supply rails. However, base-collector capacitance introduces phase shifting and results in the need for frequency compensation while operating the output near the supply rails results in saturated output transistors and introduces signal distortion.

An example common-emitter output stage is shown in Figure 7.4. We again begin our analysis by noting that the supply voltages are of equal magnitude and opposite polarity, that the load is ground-referenced and that a transistor current source comprised of Q2, R_E and a bias voltage are present. The input voltage is referenced to $-VCC$ to simplify the analysis.

Current source Q2 provides a quiescent current denoted I_Q. The output current is then given as

$$I_o = I_Q - I_C(Q1)$$

We also have

$$V_o = I_o R_L$$

If we combine these two equations and use the diode equation to express $I_C(Q1)$ as a function of V_{BE1}, we obtain

$$V_o = -R_L \beta EXP \left(\frac{V_I}{V_T}\right) - I_Q$$

The transfer characteristic is shown in Figure 7.5.

While there are similarities to the transfer characteristic of the emitter follower circuit, there are two very important differences. The first is that the transfer characteristic for the common-emitter stage is exponential. This means that as V_I changes, the output signal V_o will exhibit

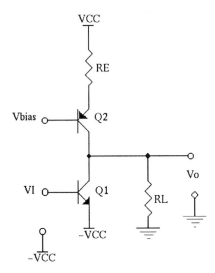

Figure 7.4   Common-emitter output stage.

distortion due to the curvature of the transfer characteristic. The linear nature of the emitter follower stage inherently provides less distortion. The second difference is that a change in V_I of only a few tens of millivolts can result in V_o traversing its entire voltage range. The emitter follower circuit requires V_I to move across its entire range in order for V_o to do so.

The maximum and minimum values of the output voltage are again dependent on the value of R_L. If R_L is large enough that $R_L I_Q > VCC$, then V_o will have values between $VCC - VCE_{sat}(Q2)$ and $-VCC + VCE_{sat}(Q1)$. If R_L is small, the maximum value of V_o will be limited to the voltage across the load resistor $V_o = R_L I_Q$.

The common-emitter output voltage signal can exhibit distortion at both peak and valley of the output waveform. Distortion from signal "clipping" will be evident if the active output transistor enters saturation, or if the maximum output voltage is clamped to $R_L I_Q$.

7.3   The Class B (Push-Pull) Output

Class A output stages are characterized by having some quiescent current flowing in the output transistors at all times. This means that power is being dissipated in the output transistors even if there is no ac input signal. This has some important consequences:

- Power dissipation raises the junction temperature of the I_C and increases the possibility that the I_C may fail.

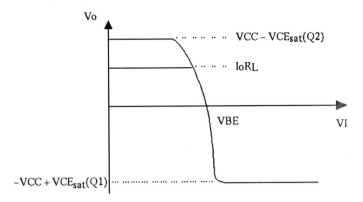

Figure 7.5   Common-emitter output stage transfer function.

- One of the considerations for sizing of integrated transistors is power dissipation. Transistors that dissipate more power must be physically bigger, and this increases die size. Larger die size leads to reduced yields because there are fewer die per wafer and more chance that defects in silicon will affect a given die. This increases the cost of manufacturing an IC and reduces the manufacturer's profit.

- When considering battery-powered ICs, wasted power translates to reduced battery life.

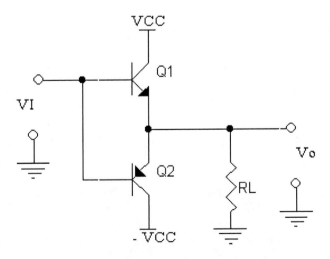

Figure 7.6   Class B output stage.

The Class B output stage addresses all these problems by dissipating no power during periods of no ac input. These circuits use two active devices to provide power to the load. Only one of the two transistors is on at a given time, each conducting for one half cycle of a sinusoidal input signal. A typical Class B output stage is shown in Figure 7.6.

Note that Q1 is an NPN transistor while Q2 is a PNP. Use of both polarities results in calling this circuit a complementary output driver. The PNP transistor is usually a vertical or substrate device. This circuit can be considered as two emitter follower stages connected in parallel.

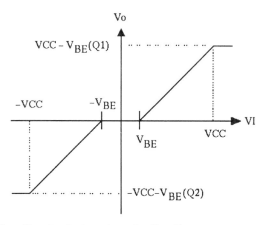

Figure 7.7   Class B output stage transfer function.

The transfer characteristic of the Class B output stage is shown in Figure 7.7. If $V_I = 0$, there is no current flowing in the load and $V_{BE1} = V_{BE2} = 0$. Both transistors are off. As V_I becomes more positive, Q1 reaches $V_{BE}(on)$ and begins to conduct. Further increases in V_I result in increases in V_o. The slope of the graph in this region is approximately 1. If we again assume V_I is generated on the I_C, we reach $V_I = VCC$ at which point $V_o = VCC - V_{BE}$. Similarly, as V_I becomes negative, Q2 reaches $V_{BE}(on)$ and conducts, eventually resulting in $V_I = -VCC$ and $V_o = -VCC - V_{BE}$.

Note the deadband of $2V_{BE}(on)$ centered at the origin. This deadband results in crossover distortion as the input voltage crosses over the origin and conduction changes from Q1 to Q2 and vice versa. This distortion may be acceptable for very wide swings in input voltage, since the deadband becomes small when compared to the amplitude of the input signal. However, there is no output signal for inputs contained in the deadband, and severe distortion for signals with amplitude slightly larger than $V_{BE}(on)$.

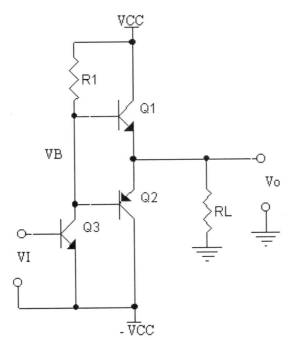

Figure 7.8   Class B output stage implementation.

Figure 7.8 shows a practical implementation of a Class B output stage. If $V_o = 0$, $V_B = 0$ and Q3 must sink the bias current through R_1. Thus,

$$I_{bias} = I_C(Q3) = \frac{VCC}{R_1}$$

The minimum value of V_o is obtained when transistor Q3 saturates. This occurs for large values of V_{BE3}.

$$V_{o(min)} = -VCC + VCEsat(Q3) + V_{BE2}$$

As V_{BE3} decreases, V_B traverses from $-VCC+V_{CEsat}(Q3)$ to $-V_{BE(on)}$. Q2 and Q3 operate in the forward active region. Q2 acts as an emitter follower and V_o follows V_B. As V_{BE3} decreases further, $I_C(Q3)$ decreases and Q1 begins to turn on. V_o now follows V_B as Q1 becomes the emitter follower. The maximum value of V_o is reached when Q3 is cut off:

$$V_{o(max)} = VCC - V_{BE1} - I_B(Q1)R_1$$

For large values of β, we also have

$$V_{o(max)} = I_C(Q1)R_L = \beta I_B(Q1)R_L$$

Substituting and rearranging leads to

$$V_{o(max)} = \frac{VCC - V_{BE1}}{1 + \frac{R_1}{\beta R_L}}$$

7.4 The Class AB Output Stage

The Class B stage would be nearly ideal if the crossover distortion were not present. The Class AB stage eliminates crossover distortion by causing both output devices to conduct a small quiescent current when $V_I = 0$.

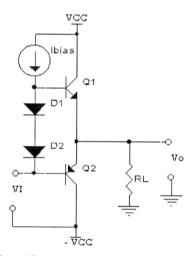

Figure 7.9 Class AB output stage.

Current source Ibias forces a current to exist in diodes D1 and D2. Since the base-emitter junctions of these diodes are in parallel with those of Q1 and Q2, the transistors are also forced to conduct. A typical characteristic for this circuit is shown in Figure 7.10. The deadband has been eliminated.

7.5 CMOS Output Stages

In general, everything we have learned regarding bipolar output stages is true for CMOS as well. However, there is the issue of CMOS processing's low transconductance to consider. Either very large devices or very large values of VGS are required for currents in the tens of milliamps range. One possibility is to use composite biCMOS devices as the output driver

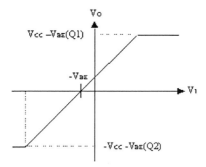

Figure 7.10 Transfer function for the Class AB output stage shown in Figure 7.9.

transistors. An example of a CMOS Class AB output stage is shown below in Figure 7.11.

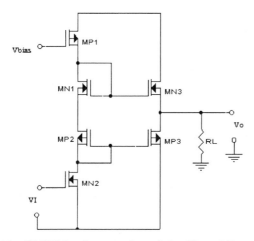

Figure 7.11 CMOS implementation of the Class AB output stage.

7.6 Overcurrent Protection

Many types of protection circuitry can be included in an integrated circuit, but one feature that is nearly universal is overcurrent protection. Overcurrent protection usually takes the form of short circuit protection, operating current limit or both. It is common to have short circuit current limit be less than the operating current limit. This is called current limit foldback. Many circuits also include a thermal shutdown circuit in addition to the current limit circuit. The thermal sensing elements are physically located next to or within the output transistors. In an overcurrent condition, current limit circuitry acts first to limit output current, thus limiting the on-chip power dissipation to a survivable level. As the output stage dissipates power, the die heats up and the thermal shutdown circuit becomes active. This usually results in turning the output stage completely off. Thermal shutdown hysteresis is usually built in, and thus the die temperature must decrease to a lower level in order for the output stage to turn back on. In this manner, the IC is protected from output fault conditions that could otherwise destroy the IC.

One of the simplest methods of current limit is shown in Figure 7.12. In this circuit, resistor R_{SENSE} measures output current. When the voltage drop across this resistor is $V_{BE(on)}$, transistor Q3 will turn on and steal base drive away from output transistor Q1. The extra base drive is shunted to the load with the result that output current is approximately limited to

$$I_{LIM} = \frac{V_{BE(on)}}{R_{SENSE}}$$

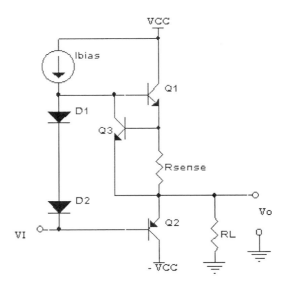

Figure 7.12   Class AB output stage with output protection provided by Q3.

As output current increases and the current limit circuit becomes active, further increase in the output loading will result in a decrease in the value of V_o. However, this current limit circuit functions properly even with V_o shorted to ground (equivalent to $R_L = 0$). This circuit takes up very little die area, but it has several drawbacks:

- The requirement that a $V_{BE(on)}$ exist across R_{SENSE} results in a decrease in efficiency and an increase in the minimum voltage between VCC and V_o. This will have important consequences for battery operated systems.

- The current limit value depends on the absolute values of both V_{BE} and R_{SENSE}. Wide variation can be expected in both parameters, and so specification limits will be wide to accommodate this variation.

- Integrated resistors usually have a positive temperature coefficient. V_{BE} decreases with respect to temperature. The result will be a strong negative temperature coefficient associated with the value of current limit. This is usually a good thing, since lowering the current limit will decrease power consumption and lower the heat generated on-chip. It is important to realize that increasing the current limit with temperature can lead to thermal runaway, where the current limit may stop working altogether.

- This circuit only protects the output stage while transistor Q1 is sourcing current to the load. There is no protection in the event that $-VCC$ is shorted to ground through Q2.

Many variations exist on this theme along with many patents for clever circuits that address some of the issues above.

7.7 Chapter Exercises

1. Use the emitter follower circuit in Figure 7.1. $VCC = 5V$, $V_{bias} = -4.1V$, $R_E = 20K\Omega$. Draw the transfer characteristic if $R_L = 100K\Omega$. Assume $V_{CE(sat)} = 200mV$.

2. Repeat exercise 1 with $R_L = 50K\Omega$ and $20K\Omega$. Find the minimum value of R_L for which no clipping of the output signal occurs.

3. Use the common-emitter circuit in Figure 7.4. $VCC = 5V$, $V_{bias} = 4.4V$, $R_E = 100\Omega$. Find the minimum value of R_L for which no clipping occurs.

4. For the circuit defined in exercise 3, draw the transfer characteristic for $-4.5V < V_I < -4.2V$. $I_S = 200E - 18A$ for Q1.

5. Use the push-pull output stage shown in Figure 7.6 with $VCC = 5V$ and $R_L = 10K\Omega$. Draw the transfer characteristic.

6. For the circuit in exercise 5, draw the waveforms for V_o, $I_C(Q1)$, $I_C(Q2)$ and I_{OUT} if V_I is a sinusoid with amplitude of 1V (zero to peak).

7. Repeat exercise 6 for a sinusoidal input of 3V (zero to peak).

8. Repeat exercise 6 for a sinusoidal input of 5V (zero to peak).

9. Use the circuit in Figure 7.8 with $R_L = 20K\Omega$, $R_1 = 5K\Omega$ and $VCC = 5V$. Let $\beta_{NPN} = \beta_{PNP} = 100$ and $I_{S(NPN)} = 200E - 18A$. What are $V_{o(max)}$ and $V_{o(min)}$? What is the minimum value of R_L before the output voltage is clipped by the resistor value? Plot the transfer characteristic for $-4.5V < V_I < -4.2V$.

References

[1] Baker, R. Jacob, et al., *CMOS Circuit Design, Layout and Simulation*, IEEE Press, New York, c. 1998.

[2] Gray, Paul R., and Mayer, Robert G., *Analysis and Design of Analog Integrated Circuits*, 2nd edition, John Wiley and Sons, Inc., New York, c. 1984.

[3] Millman, Jacob, and Grabel, Arvin, *Microelectronics*, 2nd edition, McGraw-Hill Book Company, New York, c. 1987.

[4] Moser, Jay D., *ELE536 Class Notes: Amplifier Gain and Output Buffer Stages*, Cherry Semiconductor Corporation Training Memorandum, 1997.

chapter 8

Pitfalls

This chapter illustrates some commonly made design errors. The case studies describe actual circuits that failed to function as specified after fabrication. This required redesign and refabrication, a costly time consuming process. Many times, market opportunities pass before circuits can be fixed.

8.1 IR Drops

Voltage drops in reference and power supply lines have a dramatic effect on voltage references and comparators. Error voltages get amplified and cause circuits to fail.

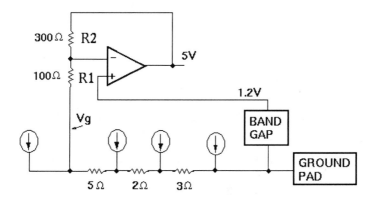

Figure 8.1 Small resistances in the ground line cause a shift in the voltage V_g. This shift is amplified by the opamp, in this case, by a factor of 3.

Case 1 Ground Line Drops

The circuit shown in Figure 8.1 is designed to be a 5 V reference. The bandgap voltage of about 1.2 V is amplified by a factor of 4 to produce the desired 5 V output. However, small resistances in the ground line, together with sometimes large ground line currents from a variety of sources produce a shift in the opamp output voltage. A 10 mA ground line current results in a 100 mV drop V_g. This is amplified by the opamp and results in a 0.3 Volt decrease in the regulator output. The opamp output is

$$V_o = V_{bg}\left(R_1 + \frac{R_2}{R_1}\right) - V_g\frac{R_2}{R_1} \tag{8.1}$$

where in Figure 8.1, R_1, R_2, V_{bg} and V_o are 100 Ohms, 300 Ohms, 1.2 V and 5 V, respectively.

Remedies

- Adding a Kelvin line, as shown in Figure 8.2, by-passes the ground line and eliminates the effect of ground line drops on the output voltage.

- Placing the bandgap, voltage divider and the opamp close to each other reduces ground and reference line drops.

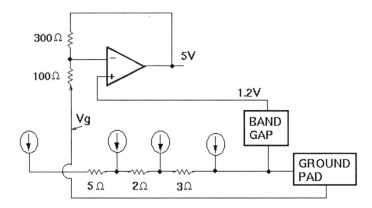

Figure 8.2   An additional low current wire (Kelvin line) to the ground pad by-passes ground line drops and eliminates the effect of ground line currents on V_g.

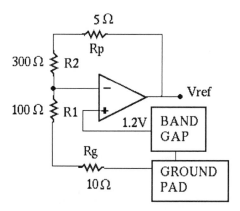

Figure 8.3 Voltage reference circuit showing parasitic ground line and reference line resistances. If $R_p/R_g = R_2/R_1$, reference and ground line resistances balance out. For the values shown, $V_{ref} = 4.53V$. The expected output is 4.8 V.

Case 2 Kelvin Line Resistance

If the voltage divider is some distance from the opamp and the bandgap, ground and reference line resistances can be significant. Problems occur even when ground line currents from other sources are not present. Figure 8.3 is a circuit containing resistances R_g and R_p representing the ground line and reference line resistances. The output voltage for this circuit is

$$V_{ref} = V_{bg} \left(1 + R_2 + \frac{R_p}{R_1 + R_g} \right) \tag{8.2}$$

If

$$\frac{R_p}{R_g} = \frac{R_2}{R_1} \tag{8.3}$$

Then

$$V_o = V_{bg} \left(1 + \frac{R_2}{R_1} \right) \tag{8.4}$$

Ground line resistance has been balanced by resistance in the reference line. For an accurate output, currents from other sources should not flow in the ground and reference lines.

Remedies

- Increase feedback divider resistance to reduce the significance of parasitic resistances.

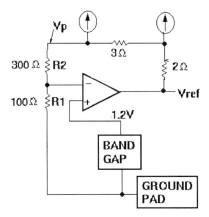

Figure 8.4   Remote loads connected to the voltage reference output produce ir drops in the reference line and result in an off set of the output, V_{ref}.

- Locate the divider closer to the reference voltage and the ground voltage.

Case 3 Reference Line Drops

Voltage reference outputs can be used in many points in a circuit. If care is not taken this can result in troublesome reference line drops as shown in Figure 8.4. The desired output V_{ref} is 4 times V_{bg}. With reference line drops, the desired output voltage appears at V_p. The output V_{ref} is offset from the desired voltage by the reference line drops.

Remedies

- A separate wire carrying little or no current, called a Kelvin line, should be used for the reference voltage. This minimizes voltage drops by eliminating load currents from the reference line.

- Layout the voltage divider close to the opamp to reduce reference line resistance.

8.1.1   The Effect of IR Drops on Current Mirrors

Small voltage drops in power supply lines have a dramatic effect on current mirrors. Output current varies exponentially with voltage drops in the supply rail.

Two identical transistors with the same base to emitter voltage will carry the same current. The second transistor mirrors the current in the first when their bases and emitters are connected. A problem arises when the second transistor is not located close to the first. A small

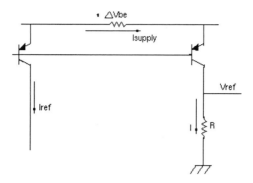

Figure 8.5   The current I does not mirror I_{ref} due to the voltage drop, delta V_{be}, in the supply rail.

ir drop in the supply rail, as shown in Figure 8.5, can cause a large difference in the transistor currents. If, for example, there is a 6 mV drop between the emitters, currents will differ by a factor of 1.26. While it is usually less of an issue, a voltage difference between bases will have the same result. Transistor collector current is given by equation 1.47, repeated here for convenience:

$$I_c = I_s exp \left(\frac{V_{be}}{V_T} \right)$$

where I_s, the saturation current, is a constant. V_{be} is the base-emitter voltage and V_T is the thermal voltage KT/q. $V_T = 25.8mV$ at room temperature. The ratio of currents in identical transistors where $I_{s1} = I_{s2}$, is

$$\frac{I_{c1}}{I_{c2}} = exp \left(\frac{\Delta V_{be}}{V_T} \right) \tag{8.5}$$

where $\Delta V_{be} = V_{be1} - V_{be2}$. When $\Delta V_{be} = 18mV$, $I_{c1}/I_{c2} = 2$.

In one application a circuit failed because the current source was used to generate a reference voltage for a comparator. Two Volts were expected but only 1.6 V occurred. This was due to a 6 mV drop in the supply rail. Rails may be connected to a variety of current sinks. Unexpected currents can flow, producing small voltage drops that result in large variations in transistor currents.

Remedies

- Transistors used as current mirrors should be located close to the transistor carrying the reference current. It is better to run the collector metal to its destination.

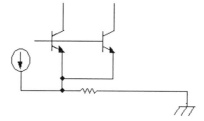

Figure 8.6 A separate wire connects common emitters, by-passing the ir drop in the supply rail.

- Power supplies should be solid between a current mirror and its reference. No under-passes should be used and wires should be thick.

- A separate wire by-passing the supply rail, as shown in Figure 8.2, can be used to connect common emitters, provided the drop in that wire is insignificant. The distance from each emitter to the rail should be the same and they should be connected to the rail at a common point.

- Emitter resistors can be used to stabilize collector current.

The use of emitter resistors in the presence of a voltage disturbance V_d is illustrated in Figure 8.7. The emitter resistor provides feedback. An increase in I_c causes an increase in the drop across R_e and results in a decrease in V_{be}. Which, in turn, decreases I_c.

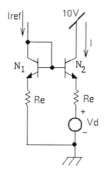

Figure 8.7 Adding emitter resistors R_e reduces output current variations in response to ground noise.

Figure 8.8 shows the effect of adding R_e to the current mirror. When R_e is zero, an 18 mV disturbance causes I_c to double from the reference

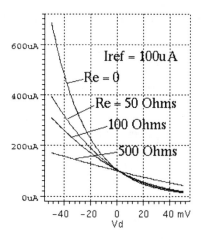

Figure 8.8   Simulation showing the improved stability as the emitter resistance increases. I_c is 100 μA. When R_e is 500 Ohms, the DC voltage across R_e is 50 mV, about $2V_T$. An 18 mV disturbance produces a 25% change in I_c.

value of 100 μA to 200 μA, (or down to 50 μA for positive V_d). As R_e is increased, changes in the voltage across it dominates over changes in V_{be}, and I_c varies linearly with V_d. Since I_c is 100 μA, when R_e is 500 Ohms, the voltage across it is 50 mV.

Current stability is a function of the number of thermal voltages (26 mV) across R_e. This voltage is

$$V = I_c R_e$$

The transistor current is a function of the base-emitter voltage

$$I_c = I_s exp\left(\frac{V_{be}}{V_T}\right)$$

With an emitter resistor present, and the change in I_c is small, small signal analysis can be used. The incremental resistance of the base-emitter junction is

$$R_{be} = \frac{dV_{be}}{dI_c} = \frac{V_T}{I_c}$$

$$R_{be} = \frac{1}{g_m}$$

where $g_m = I_c/V_T$ is the transconductance.

If the drop across R_e is one thermal voltage, then $R_e = V_T/I_c = R_{be}$. A disturbance to the supply rail, like V_d, divides between R_e and R_{be}. Any voltage across R_{be}, the base-emitter junction, contributes to the exponential change in collector current. When the voltage across R_e is V_T, $R_{be} = R_e$. In this case, any disturbance V_d divides equally between R_e and R_{be}. If R_e is larger, it will absorb more of the disturbance and the change in I_c will be less and more linear.

When R_e is 500 Ohms, the DC voltage across it is 50 mV, about $2V_T$. That means R_e is $2R_{be}$. Only one-third of the disturbance will contribute to V_{be}. Therefore, an 18 mV disturbance only produces a 6 mV change in V_{be}, so the current changes by a factor of 1.25. In this example, that's from 100 μA to 125 μA, as shown in Figure 8.8.

8.2   Lateral pnp

In the classic bipolar fabrication process, the process is tuned to produce a good npn transistor. The lateral pnp transistor is fabricated using diffusions optimized for the npn transistor. Lateral pnp layout is shown in Figure 8.1. The p-type npn base diffusion is used as both the emitter and collector of the lateral pnp. The pnp base is the n-type epitaxial layer. Modern processes have additional steps that permit a true vertical pnp to be fabricated. In spite of some performance sacrifices, lateral pnps are attractive because they can be fabricated with fewer process steps. The lateral pnp is a widely used transistor.

8.2.1   The Saturation of Lateral pnp Transistors

Problems can arise when a lateral pnp goes into saturation. Collector current reverses due to a parasitic pnp turning on and dumping current to the substrate. Beta drops to zero and goes negative.

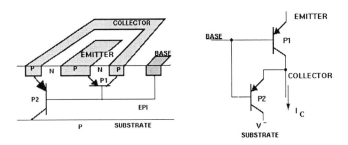

Figure 8.9   The lateral pnp transistor P1 is accompanied by an undesired parasitic pnp P2.

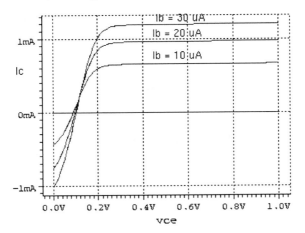

Figure 8.10 Simulated lateral pnp including the parasitic pnp causing I_c to reverse in saturation.

Lateral pnp transistors behave in unexpected ways when they go into saturation. This is due to the parasitic transistor P_2 shown in Figure 8.9. It turns on when the lateral transistor P_1 goes into saturation. With the lateral transistor in saturation, the collector voltage approaches the emitter voltage, forward biasing the parasitic transistor base-emitter junction on at one V_{be}.

When the parasitic transistor turns on, current flows through it from the P_1 collector to the substrate. In normal operation, collector current is due to the lateral pnp P_1. It flows out of the collector contact. However, as can be seen in the characteristic curves shown in Figure 8.10, as P_1 goes into saturation, its collector current reverses. This is due to the current pulled into the collector contact and dumped into the substrate by the parasitic transistor. Base current continues to flow. Beta, defined as I_c/I_b, drops to zero and becomes negative.

Single transistor SPICE models do not predict the behavior of lateral pnp transistors in saturation. The negative currents are due to the parasitic transistor P_2 shown in Figure 8.9.

8.2.2 Low Beta in Large Area Lateral pnps

Very low betas have been observed in large lateral pnp transistors. Beta drops dramatically as pnp size increases due to recombination in the buried layer. Theory indicates beta is proportional to the emitter area divided by the area of the emitter perimeter.

A graph of beta as a function of pnp transistor size is shown in Figure 8.11. Beta has been observed to vary inversely with emitter area. 10 μm x 10 μm transistors have a measured beta of 70-80. 26 μm x 26 μm transistor has a measured beta of 10-15[1].

Glaser and Subak-Sharpe [2] present a theory. The drop off in beta is attributed to recombination in the buried layer. The theory shows beta proportional to A_L/A_V, where A_L is the emitter perimeter multiplied by the junction depth, and A_V is the drawn emitter area.

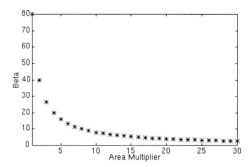

Figure 8.11   Beta vs area multiplier approximates data taken from several lots of regulators. Beta is 80 when the area multiplier is one.

Transistor performance can be thought of in terms in the holes emitted from the emitter. Holes from the emitter sidewalls diffuse to the collector where they contribute to collector current. Holes emitted from the emitter bottom, area A_V, have a high probability of diffusing to the buried layer where the recombination rate is high. Each hole that recombines requires an electron supplied by base current. There will be other holes that either make it through the buried layer or diffuse to the isolation wells. These are swept across the reversed biased junction and contribute to substrate current.

If you assume holes emitted from the bottom of the emitter recombine in the buried layer and are the main component of I_b, and holes emitted from the emitter sidewalls contribute to collector current, then beta, the ratio if I_c to I_b, is proportional to the ratio of the sidewall area to the bottom area.

Because of the doping difference between the epi and the buried layer there is a built-in potential that reflects holes from the buried layer. Quoting Glaser and Subak-Sharpe [2, Sec 2.17, p. 66]:

> Actually, if the emitter stripe width is not significantly greater
> than twice the distance from the bottom of the emitter to

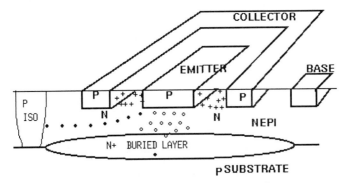

Figure 8.12   The lateral pnp is shown.

the buried layer, most of the holes injected from the bottom face of the emitter will be reflected from the buried layer. If, in addition, the width of the collector region is very wide, most of these holes will be collected.

8.3   npn Transistors

The higher mobility of electrons compared to holes makes the npn easier to make than the pnp. Bipolar fabrication processes favor the npn. In general, npns have higher beta, lower leakage, larger early voltage, and can carry higher currents than pnps of the same size. This section illustrates two design problems that can arise in saturation and at elevated temperatures.

8.3.1   Saturating npn Steals Base Current

The structure of integrated circuit npn transistors produces an unwanted parasitic pnp. When the npn saturates, the pnp turns on producing unexpected results.

It is common to construct a number of current sources with bases connected to a common point. If one npn saturates, it turns on a parasitic pnp that sinks base current to the substrate. This turns off the other npn current sources causing circuits to fail.

An example of this failure mode is shown in Figure 8.13. When the pnp transistor P_1 is cut off, N_4 saturates, causing the output V_o to go low. However, the parasitic pnp in N_4 robs base current from remaining transistors. The circuit fails because the bias current I_3 is turned off.

The parasitic pnp transistor is shown in Figure 8.14. As the npn goes

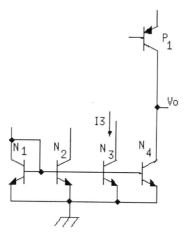

Figure 8.13   Current sources with bases connected.

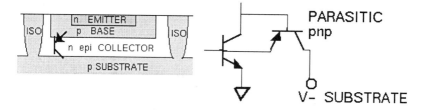

Figure 8.14   The npn base, collector and substrate form a parasitic pnp.

into saturation, the npn collector voltage approaches its emitter voltage. The npn base-emitter junction is forward biased, and with the collector voltage approaching the emitter voltage, the pnp base collector junction also becomes forward biased. This turns on the parasitic pnp.

Remedy

Placing a 1K resistor in series with the base of the saturating transistor has been found to remedy the failure.

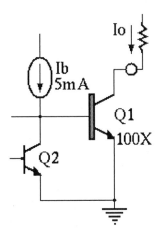

Figure 8.15   The power transistor Q1 is held "off" by Q2. However, at elevated temperatures, Q1 leakage current is excessive, turning on Q1, causing the output current I_o to exceed specifications.

8.3.2   Temperature Turns On Transistors

Collector current increases with temperature. At low V_{be} ($250mV$), a collector current of a few nanoamperes is observed at room temperature. But at elevated temperatures ($135°C$), collector currents in the hundreds of microamp range flow, causing circuit failure in spite of the low V_{be}. This is due to the exponential dependence of saturation current I_s on temperature.

The transistor Q2 in Figure 8.15 controls the power transistor Q1. When Q2 is on, it sinks Q1's base current, holding Q1 off. The saturation resistance of Q2 is 50 Ohms. When sinking $5mA$, its V_{ce} is 0.25 V. At room temperature this holds Q1 off. Leakage current of about $6nA$ flows in the off Q1. At elevated temperatures, the saturation resistance of Q2 increases, but the current through it, I_b, may decrease. Here we assume the voltage across Q2 does not change appreciably with temperature. It remains at 0.25 V. In spite of this low V_{be}, Q2 begins to turn on at elevated temperatures.

At elevated temperatures, the saturation current I_s increases causing the Q1 collector current I_o to increase from nanoamps to hundreds of microamps. Since Q1 is "off," this constitutes circuit failure.

Since $I_c = I_s exp(V_{be}/V_T)$, where V_T is the thermal voltage, at room temperature Q1 carries $100mA$ at $V_{be} = 0.68V$. This corresponds to $I_s = $ 4E-13 A. With $V_{be} = 0.25V$, the collector current for Q1 is $6.2nA$.

The saturation current I_s is a function of the strongly temperature dependent quantity, intrinsic carrier concentration, n_i. SPICE models the temperature dependence of the saturation current I_s using the following

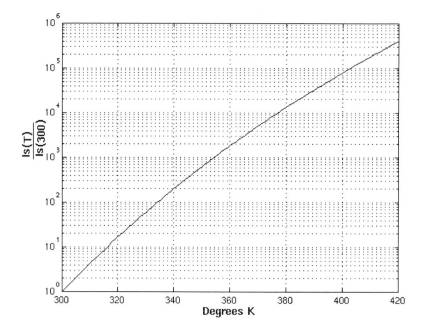

Figure 8.16   A SPICE simulation showing Is is a nearly exponential function of temperature.

equation:

$$I_s(T_2) = I_s(T_1) \left(\frac{T_2}{T_1}\right)^{XTI} exp\left[-\frac{qE_g}{KT_2}\left(1 - \frac{T_2}{T_1}\right)\right]$$

- $T_1 = 300°K$.

- $T_2 = 415°K = 135°C$. The junction temperature is 10 degrees above the $125°C$ ambient.

- $I_s(T_1) = $ 4E-13.

- The SPICE parameter (I_s temperature effect exponent) XTI = 1.7.

- The thermal voltage $KT = 0.0259V$ at $T = T_1$ (room temperature).

- The bandgap voltage $E_g = 1.12V$.

At $T_2 = 415°K$, I_s has increased by a factor of 2.8E5 above the room temperature value to 0.11 μA. With V_{be} held constant at 0.25 V, I_c increases to 119 μA. more than one-tenth of a milliamp. This represents a failure since V_{be} is only 0.25 V, the transistor Q1 is designed to be OFF.

Remedy

The transistor Q2 has to be large enough to handle the leakage current from Q1 at elevated temperatures.

8.4   Comparators

This section discusses three failure modes for comparators. The first is "headroom" failure, where there is not enough voltage across the transistor providing the bias current. The transistor saturates causing the circuit to fail. In the second case, the allowable range of input voltages is exceeded. The third is a case where charge stored in a Darlington input causes an erroneous comparison.

8.4.1   Headroom Failure

Comparator tail current is cut off due to insufficient voltage across the current source. The two comparator modes look OK, but switching from a LOW output to a HIGH output fails.

Example 1

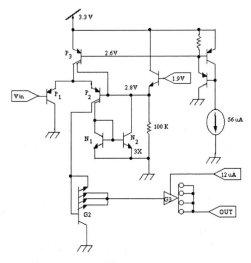

Figure 8.17   Logic level comparator.

The circuit shown in Figure 8.17 is designed to act as a logic level input comparator. A LOW input turns P_1 on and P_2 off. With P_2 off, current to the current mirror G_2 is zero. This represents a HIGH to the I2L gate G_3. The output is LOW. When the input is HIGH, P_1 is off, P_2 is on, and the output is HIGH. Hysteresis is achieved by the current mirror N_1 and N_2.

With P_2 on, N_1 and N_2 turn on. N_2 pulls the base of P_2 to one V_{be} below the reference voltage of 1.9 V. That's about 1.2 V. This low base voltage snaps P_2 to fully on. The circuit is shown with a LOW input.

Trouble occurs because there is not enough headroom. When the voltage across P_3 is low, P_3 saturates, current decreases, and the comparator fails.

With a zero input voltage P_1 is on. P_2, N_1, and N_2 are off. 28 μA flowing through the 100 K resistor from the current mirror P_3 drives the base of P_2 to 2.8 volts. The emitters of P_1 and P_2 are at one V_{be} (0.7 V at room temperature). The base of P_3 is one V_{be} below VCC. That's about 2.6 V.

When the input goes HIGH, P_1 turns off. The emitters of P_1 and P_2 attempt to rise to one V_{be} above the base of P_2. That's 2.8 + 0.7 = 3.5 V at room temperature. However, there is not sufficient voltage across P_3 to maintain current. With no current in P_2, N_1, and N_2 are off. N_2 fails to pull the base of P_2 low. P_2 stays off. The output remains LOW. The circuit fails to recognize a HIGH input.

The problem is worse at high temperatures because the 100 K resistor resistance increases.

Example 2

Consider the comparator with hysteresis shown below. With a LOW input, P_2 and N_1 are off. The current source turns P_4 on and V_n is one V_{be} above V_{ref}. When P_2 is on, N_1 is also on. N_1 sinks the current source and pulls current from N_2, turning it on and pulling V_n one V_{be} below the reference. This gives a hysteresis of $2V_{be}$.

Trouble occurs because there is not enough headroom. When the voltage across P_3 is low, P_3 saturates, current decreases and the comparator fails.

When V_{in} goes high, P_1 turns off, the emitters of P_1 and P_2 attempt to rise to one V_{be} above V_n to turn P_2 on. However, with a small voltage across P_3, it saturates and no current flows to P_2, N_1 remains off. The comparator fails.

The problem is worse at low temperatures where V_{be} can equal 0.8 or 0.9 volts.

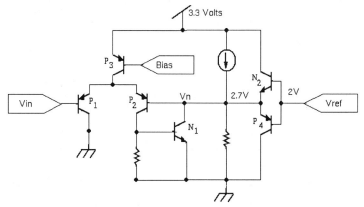

The circuit is shown with Vin LOW

Figure 8.18 As in example 1, the comparator is unable to switch when the input goes from LOW to HIGH. With a LOW input, P_1 is on, P_2 and N_1 are off. $V_n = V_{ref} + V_{be}$, about 2.7 volts for this example.

8.4.2 Comparator Fails When Its Low Input Limit Is Exceeded

In this case the comparator input voltage range is exceeded. The problem is compounded by the fact that SPICE models for transistors in saturation are poor.

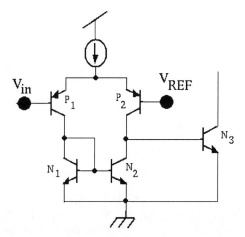

Figure 8.19 A circuit that fails when the input goes much below V_{be}.

Example 1

Consider the comparator shown in Figure 8.21. The minimum input voltage must be large enough to keep N_1 turned on and P_1 operating in the normal region. This requires 0.7 V across N_1 and a zero base to collector voltage for P_1. Therefore, the minimum input voltage is equal to one $V_{be} = 0.7$ V at room temperature.

The problem occurs when the voltage on the base of P_1 is too low. Even with P_1 saturated, the N_1 base voltage is not high enough to turn N_1 on and the circuit fails.

The circuit in Figure 8.17 fails if the input is grounded. This is outside the input voltage range. Consider the case where V_{REF} is a positive voltage, say 2 V. With the input grounded, one would expect P_1 to be on and P_2 to be off. However, the low input voltage at the base of P_1 does not provide enough voltage to keep N_1 on. With N_1 off, N_2 is also off. This allows leakage current from P_2 to turn N_3 on. The comparator fails to function properly.

The problem also occurs in the complimentary circuit where the input transistors are npn input transistors. In that case the input voltage can not equal the positive rail, but should be one V_{be} below it.

One remedy is to use a Darlington input.

Example 2

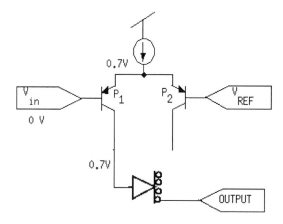

Figure 8.20   This comparator is designed to have a LOW output when the input is LOW. With a LOW input, P_1 turns on and provides current to the I2L gate. This represents a HIGH input to the I2L inverter.

The comparator in Figure 8.20 fails when the input goes LOW and P_1 attempts to turn on. The emitter is one V_{be} above the base, about 0.7 V at room temperature. The collector tries to go to the one V_{be} needed to turn the I2L gate on. This leaves zero volts across P_1. With zero volts across P_1, no current flows and the HIGH input to the I2L gate is not achieved.

As in example 1 above, the remedy is to use a Darlington input. With a diode in series with the input, the base and emitter are raised by one V_{be}. When the input is zero volts, the emitter of P_1 will be at $2V_{be}$. The collector is at the I2L HIGH of one V_{be}. That leaves one V_{be} across P_1 and ensures sufficient current to turn the I2L gate on.

8.4.3   Premature Switching

A circuit using a comparator designed to generate a delay failed. Charge stored on a floating node caused the comparator to switch prematurely. The circuit failed to generate the expected delay.

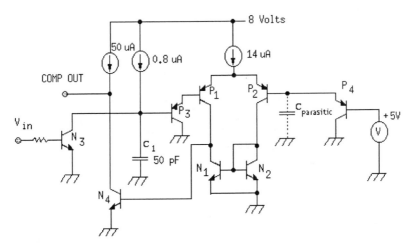

Figure 8.21   Delay circuit fails because the base of P_2 floats.

The circuit shown in Figure 8.21 is designed to produce a delay equal to the amount of time it takes the capacitor to charge up to 5 V. The output is designed to go high a fixed time after the input goes low. With a high input, N_3, P_3 and P_1 are conducting. The emitters of P_1 and P_2 are at about 2 V_{be} plus the saturation voltage of N_3. This is about 1.4 V at room temperature. P_2 and P_4 are off. Due to collector-emitter leakage in P_4, the base of P_2 will discharge to a small V_{be} below its emitter, or about 0.9 V. When the input goes low, N_3 turns off and the capacitor begins to charge. When the comparator operates properly, the base of P_2 is charged by P_2's small base current until it reaches 5.7 V, and P_4 turns on. This causes the comparator to switch. However, if the current gain, beta, of P_2 is large and the capacitor slews quickly, a larger base current is needed to charge the base of P_2. This causes a large enough collector current to flip the comparator prior to P_4 turning on. Thus, the proper delay is not achieved.

Remedy Number 1

The floating base of P_2 can be charged with a portion of its collector current, instead of just its base current, by splitting the P_2 collector and tying one collector back to the base as shown in Figure 8.22. When the capacitor is slewing positive, the collector tied to the base of P_2 charges the base from 0.9 V to 5.7 V as before, but the current in the other P_2 collector is never large enough to prematurely flip the comparator. The comparator only flips when P_4 turns on as the base of P_2 reaches 5.7 V, as expected.

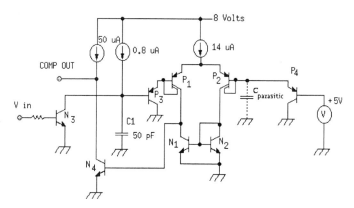

Figure 8.22   Failure corrected by tying a collector of P_2 back to its base.

Remedy Number 2

The floating base of P_2 can be charged to the proper voltage using a small current source as shown in Figure 8.23. The small current holds the base of P_2 one V_{be} above the reference input voltage. That's approximately 5.7 V at room temperature.

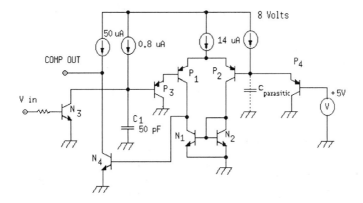

Figure 8.23   A small current turns on the base-emitter diode of P_4 and clamps the base of P_2 one V_{be} above the base of P_4.

8.5 Latchup

Parasitic transistors turn on producing a low resistance path between power rails. Large currents flow causing thermal destruction. Processing, layout, and circuit design techniques, properly applied make latchup unlikely. The structure of CMOS creates parasitic transistors that can cause latchup. Bipolar circuits can also latchup; examples are included in this section.

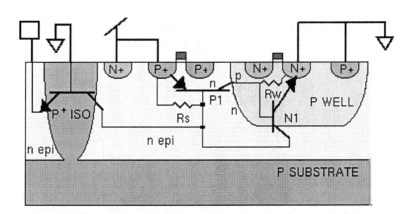

Figure 8.24 Physical source.

A representation of CMOS structure is shown in Figure 8.24. PMOS transistors are placed in the n epi. NMOS transistors are in a pwell in the n epi. The two parasitic transistors in the pwell-epi area are structured so that if one turns on it tends to turn the other on. They form the silicon controlled rectifier (SCR) structure shown in Figure 8.25. Once turned on they stay on and form a low resistance path between Vdd and ground. The parasitic transistor formed by the p+ ISO well can act as a trigger.

The n-type epi is connected to the positive supply, and the pwell is connected to the negative supply. This reverse biases junctions and isolates MOS transistors. If the epi voltage drops one V_{be} below the positive supply, the parasitic pnp is turned on. Similarly, if the pwell rises one V_{be} above the negative supply, the parasitic npn turns on.

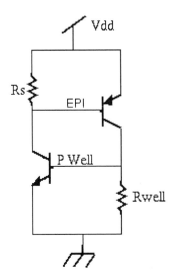

Figure 8.25   Parasitic transistors form an SCR structure. If the voltage across R_{well} or R_s exceeds one V_{be}, latchup is triggered.

Latchup Triggers

- Latchup was observed in a controller IC. When any pin was pulled 700 mV below the negative supply, latchup occurred. In Figure 8.24, the epi tub on the left is connected to a pad. A buried layer and the epi form a diode with the ISO that acts as ESD protection for the pad. This junction is shown as the base-emitter junction for the parasitic ISO npn transistor in Figure 8.25. When the pad is pulled one V_{be} (700mV) below the negative supply, the parasitic transistor turns on. This pulls current from the adjacent epi tub containing the CMOS transistors. Currents flowing through the epi reduce the epi voltage one V_{be} below the positive supply, triggering latchup.

- Over driving drains of PMOS or NMOS transistors triggers latchup. If the drain of a PMOS transistor is raised one V_{be} above the positive supply, the drain epi pn junction becomes forward biased, causing currents to flow in the epi, triggering latchup.

  Similarly, if the drain of an NMOS transistor is pulled one V_{be} below the negative supply, the drain pwell pn junction becomes forward biased, causing currents in the pwell. This can trigger latchup.

- Power supply transients can trigger latchup. Power supply transients that forward bias the epi-drain junction on the PMOS transistors or the pwell-drain junction on the NMOS transistors can trigger latchup. Also, when the power is turned on, the pwell-epi parasitic capacitance is initially uncharged. If the power is turned on too fast, this capacitor remains uncharged and the supply voltage appears across R_s and R_{well}, since the pwell and epi are shorted by their uncharged parasitic capacitance.

Remedies

A number of design techniques are used to reduce the probability of latchup.

- Generous use of epi tub and pwell bias contacts. Placement of bias contacts between PMOS transistors and NMOS transistors decreases the values of parasitic resistances.

- Reduce epi tub resistance R_s by contacting the buried layer with a deep N diffusion.

- Improve supply busing. Proper reverse biasing of the epi tub and the pwell requires that they be connected to the most positive supply and the most negative supply, respectively. Supply under passing and serpentine routing should be avoided. Supply lines should be wider than minimum to reduce voltage drops.

- Isolate MOS transistors connected to bond pads. Latchup requires NMOS and PMOS transistors to be in close proximity. If transistor drains are connected to pads and the pads are driven beyond the supply rails, parasitic transistors may be turned on, triggering latchup. Separating PMOS and NMOS transistors increases the base width of the parasitic lateral pnp. This decreases its beta

and reduces the loop gain of the SCR structure below the value
necessary to sustain latchup.

- Use guard rings between NMOS and PMOS devices. These rings
  are made of N+ source/drain diffusions and are connected to the
  positive supply rail. They reduce Rs, the epi resistance, and reduce
  the beta of the lateral parasitic pnp. Regroup transistors according
  to type with a greater distance between PMOS and NMOS transistors. Avoid a "checkerboard" layout with PMOS and NMOS
  devices mixed together.

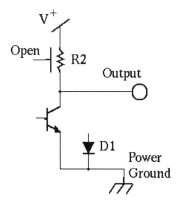

Figure 8.26   The epi tub containing R2 was left floating to prevent currents
when the output is driven above V_+. This enabled latchup.

8.5.1   Resistor ISO EPI Latchup

The pnpn structure formed by a p-type resistor in the n-type epitaxial
tub, together with p-type isolation and a second epi tub, forms a pnpn
structure that can latchup. This structure occurs in bipolar as well as
CMOS integrated circuits. The following case study illustrates how this
can happen and remedies to be taken.

If the epi is allowed to float, a power supply transient can trigger
latchup.

The resistor R2 in the output circuit shown in Figure 8.26 was placed
in a floating epi tub. The usual practice is to bias the epi tub at the high
supply voltage. This was not possible for this part because transients in
inductive loads can force the output voltage above the V_+ power supply
voltage, forward biasing the resistor epi pn junction.

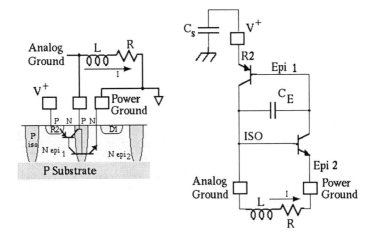

Figure 8.27   Two epi tubs separated by a p-type isolation region are shown. The p-type resistor R2, the epi tubs and the iso form a pnpn SCR structure. External wires have inductance and resistance. C_E is the EPI_1 to iso capacitance.

During testing, a transient current pulse in the analog ground caused the analog ground voltage to rise above the power ground, triggering latchup and destroying the circuit.

An SCR structure is formed by a p-region in epi tub separated from an adjacent n-type epi tub by p-type isolation. This is the pnpn SCR (silicon controlled rectifier) structure shown in Figure 8.27. The parasitic npn and pnp transistors drive each other. The collector current of each transistor provides base current for the other.

The circuit failed in testing. Figure 8.27 is a schematic representation of the parasitic SCR structure and external components. When the part was tested, the shunt power supply capacitor C_s was precharged to V^+. The analog and power ground pins were connected to the power supply using long (1 foot) wires. When the part was connected to the precharged supply capacitor C_s a large current pulse flowed from V^+, through the part, and out the analog ground. The pulse created a voltage across the inductance of the wire connecting the analog ground to the power supply. This produced a differential voltage between the analog ground and the power ground forward biasing the base-emitter junction of the parasitic npn and triggering latchup.

Connecting the precharged power supply shunt capacitor C_s to the V_s pin produced a rapid increase in the V^+ supply voltage applied to the circuit. Since the parasitic epi to iso capacitance, C_E was uncharged, the supply voltage transient appeared across the base-emitter junctions of the parasitic pnp and npn, shown in Figure 8.27, forward biasing

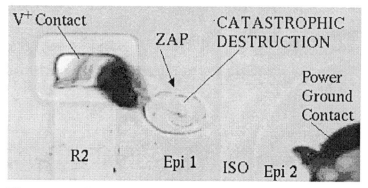

Figure 8.28   Micro photograph showing destruction.

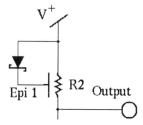

Figure 8.29   Schottky diode prevents R2 epi junction forward bias.

them. The parasitic pnp shorts V^+ to analog ground, discharging C_s. The resulting pulse of current creates a voltage in the inductive external wire connecting analog ground to the supply, turning on the npn, and completing latchup. Although latchup only lasts for the duration of the current pulse, it is a catastrophe for the silicon as shown in Figure 8.28.

Remedy

- The part failed during test. The external power supply shunt capacitor C_s was reduced. This reduced the power supply transient that triggered latchup.

- The Schottky diode shown in Figure 8.29 was added. Since it conducts at 0.4 V, it denies the parasitic pnp sufficient V_{be} to turn on. The Schottky blocks current flow in the resistor-epi *pn* junction when the output pin is driven above V^+. The Schottky diode layout is shown in Figure 8.30.

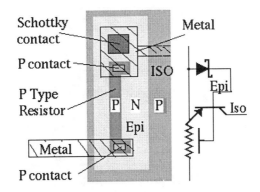

Figure 8.30 Layout for Schottky biasing of the epi tub.

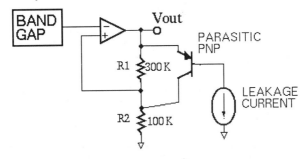

Figure 8.31 A parasitic transistor shunts current around R1 causing a drop in the output of the voltage regulator.

8.6 Floating Tubs

Unexpected currents flow in p-type resistors placed in n-type epi tubs when the tub floats. This is due to a parasitic lateral pnp transistor formed by the resistors and the n-type epi tub. Epi substrate leakage current acting as base current turns on the parasitic transistor. The effect is more pronounced for large resistance values.

A voltage regulator designed to provide 5 V uses a 1.25 bandgap source as shown in Figure 8.31. Large valued ion implant resistors, R1 and R2, were used. The circuit failed because the voltage was observed to drift downward with temperature. A $200mV$ decrease was observed at $125°C$. This is attributed to the lateral parasitic pnp formed by the p-type resistors in the n-type epi tub shown in Figure 8.32. The transistor is turned on by epi substrate leakage current. The leakage current increases with temperature and is multiplied by the transistor beta.

The effect is more pronounced when resistor values are large and at elevated temperatures where leakage currents are more larger.

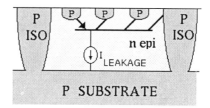

Figure 8.32 P-type resistors in the n-type epi tub form a parasitic lateral pnp transistor.

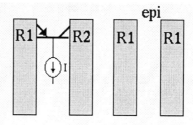

Figure 8.33 Resistors R1 and R2 are laid out in the epi tub with R2 close to the high voltage part of R1. This worsens the problem by increasing the beta of the lateral parasitic pnp.

The p-type substrate is at the lowest voltage in the circuit. P-type resistors assume circuit voltages. In this case the highest resistor voltage is 5 V. Initially current flows from the resistors to the epi tub. This charges the tub and increases the tub voltage until it is one V_{be} below the highest resistor voltage and current through the resistor-epi pn junction is cut off. The tub floats at a point where the resistor-tub pn junction is barely off. Any leakage turns it on. This pn junction is the emitter-base junction of the parasitic pnp.

The resistor layout shown in Figure 8.33 contributed to the problem by placing R2 close to the portion of R1 having the highest potential. This increased the parasitic beta.

Remedies

- The problem is avoided by biasing the tub.

- Layout the resistors to reduce the parasitic beta by increasing the emitter collector spacing. The emitter is the highest voltage portion of the resistor. The collector is another part of the resistor or another resistor at a lower potential. In the example shown here, the beta is maximized by placing R2 close to the high voltage part of R1.

- Use lower resistor values to increase current levels. Current shunted through the parasitic transistor is a small leakage current.

8.7   Parasitic MOS Transistors

Parasitic MOS transistors are formed by running metal over epi between adjacent p-type regions. These transistors can have threshold voltages in the 10 to 20 V range. They turn on when the p-type region's voltage is greater than the metal voltage by more than a threshold. Even without metal, a parasitic OSFET transistor can be formed by p-regions in epi. The gate function is performed by negative charges trapped in the oxide. These transistors produce unexpected currents causing circuits to fail.

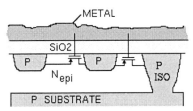

Figure 8.34   Low voltage metal can turn on parasitic MOS transistors.

The common bipolar structure shown in Figure 8.34 forms a PMOS transistor. The source and drain are p-regions in the low doped epi. The gate is metal running over the epi between the p-regions. When the gate is negative relative to the source, holes are attracted to the epi-region under the gate. When the gate voltage is below the source by one threshold voltage, the epi inverts from n to p, a channel forms in the epi, connecting the two p-regions. The voltage at which inversion occurs is determined by oxide thickness, epi doping, metal and silicon work functions, and charges trapped in the oxide.

Parasites thrive when closely spaced p-regions are separated by low doped epi. These parasites carry currents that have devastating affects on circuit performance.

The voltage of the metal relative to the epi at which mobile holes form in the epi below the metal, is called the inversion voltage. The lower the magnitude of the inversion (threshold) voltage the more easily a parasitic MOS transistor can be turned on. The fabrication process is designed to produce transistors to operate up to a voltage called the process voltage. Measurements taken on 14 V, 17 V and 30 V processes show inversion voltages generally above the process voltage. However, even in the relatively small sample of observations, inversion voltages as low as 5.23 V for the 14 V process, and 12.27 V for the 30 V process were observed.

High voltage processes are more susceptible to parasitic MOS. p-regions associated with pnp transistors are likely to be at potentials close to the high supply. Metal at low voltages (GND) is common. Larger voltage differences between p-regions and metal make high voltage processes more likely to turn on parasitic MOS structures.

Threshold voltage has a component that increases as the square root of the source to body voltage. The source is the most positive p-region. The body is the epi. Therefore, the worst case (lowest threshold voltage) occurs when the p-region and the epi are shorted.

8.7.1   Examples of Parasitic MOSFETs

There are many situations where adjacent p-regions in low doped epi lead to parasitic MOS transistors. A few are listed here.

- NPN transistor base and p-isolation

- NPN base and an underpass p-tub

- Underpass p-tub and p-isolation

- PNP collector to iso

- PNP collector to p-resistor

- P-resistor to p-resistor

8.7.2   OSFETs

IC failures can be due to a time dependent drift of circuit parameters. A circuit can be within spec during test but fail after operation of a few minutes or many hours. Drift is greater at higher temperatures and voltages. If power is removed and the IC is baked at $150°C$ for an hour and cooled to room temperature, it returns to within spec.

Drifting ICs cause difficult problems. A part can pass tests and be shipped only to fail later. The same part can drift back to within spec after being returned. The manufacturer says, "it works for us," but the customer knows better.

OSFETs are MOS transistors without the metal gate.

The function of the gate is performed by negative charge trapped in the oxide as shown in Figure 8.35. These charges attract holes to the surface under the oxide forming a channel between p-regions. OSFETs turn on when two adjacent p-type regions have a difference in voltage greater than the MOS threshold voltage. Charges trapped in the oxide cause an inversion of the n-epi surface at the oxide interface. If the n-epi

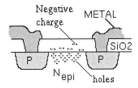

Figure 8.35   OSFET.

region is floating, it assumes a voltage approximately equal to the higher voltage p-region. This high voltage reduces the inversion threshold by the body effect. Time and temperature dependent leakage current flows between the two p-type regions *without* any metal between them. This parasite acts like a MOSFET. Since no metal is present it is called an OSFET.

The physical mechanism believed to be responsible for the drift is the movement of positive ions in the oxide under the influence of the electric field produced by the voltage difference between the two p-type regions. These positive ions have low mobility. They slowly drift away leaving fixed negative ions behind. These negative ions attract holes to the epi surface under the oxide producing the channel connecting the two p-type regions.

8.7.3   Examples of Parasitic OSFETs

Parasitic OSFETs can occur when p-regions in n-epi are closely spaced.

- PNP transistor emitter to collector

- PNP transistor collector to p-isolation

- NPN transistor base to p-isolation

- P-type (base) resistor to p-isolation

- P-type (base) resistor to P resistor

Remedies

Steps can be taken to cure bipolar IC of the affects of MOS parasites.

- Channel Stops Isolate MOS Parasites

An N+ region in the epi where a channel may form, blocks the channel by increasing the threshold voltage. For a channel to form the n-type epi must be inverted to p-type. A strongly n-type region is difficult to invert and blocks channel formation. A channel can not form in the N+ region as shown in Figure 8.36

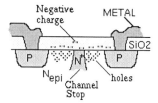

Figure 8.36   An N+ region breaks the channel continuity between the p-regions.

- Source Flapping Turns Off MOS Parasites

    Placing metal between adjacent p-regions and connecting it to a high voltage as shown in Figure 8.37, inhibits channel formation. The positive metal attracts electrons to the epi surface and counters the attraction of holes by negative charge trapped in the oxide.

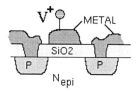

Figure 8.37   Metal at high voltage repels holes from the epi oxide interface preventing channel formation in the epi.

- Special care should be exercised if voltages exceeding the process voltage are on the chip.

8.8   Metal Over Implant Resistors

Metal over ion implant resistors forms a p-channel MOS transistor with the metal as gate. A positive voltage on the metal restricts resistor current flow.

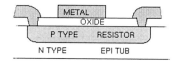

Figure 8.38   The p-type resistor, oxide and metal run form an MOS structure.

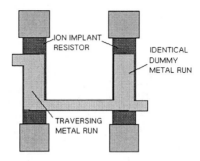

Figure 8.39   Matched ion implant resistors crossed by metal are shown (Not recommended). For accurate tracking, dummy metal is added to the second resistor.

Ion implant resistors can be matched with differences less than 0.1 percent. They are formed by ion implant doping of n-type epi tubs. This creates p-type resistive regions in the n-type tubs. The resistor is covered with oxide for isolation. If a metal run crosses the resistor, an MOS structure as shown in Figure 8.38 is formed. This structure acts like a depletion mode p-type MOS transistor. A positive voltage relative to the resistor applied to the metal repels holes and creates a depletion region in the resistor at the oxide interface. This reduces the resistor cross section and increases resistance. Current in the resistor is being pinched off. A negative voltage on the metal relative to the resistor, reduces resistivity by enhancing the hole concentration in the resistor.

A problem occurs when resistors are matched and one is crossed by a metal run. Voltage on the metal upsets the match and causes circuits to fail.

Temperature also influences resistance. This is due to an increase in lattice scattering of carriers with temperature: Resistance is approximately proportional to absolute temperature.

$$\frac{R_1}{R_2} = \frac{T_1}{T_2}$$

where temperature T is in degrees Kelvin. This represents about a 0.33% change in resistance per degree C at room temperature.

Remedies

For accurate matching of ion implant resistors:

- Avoid metal runs over resistors.

- When metal is unavoidable, matched resistors should have identical metal, as shown in Figure 8.39.

References

[1] Jay Moser, *Large Area PNPS*, CIBu Technical Note, No. JM1, Cherry Semiconductor, Jan 21, 1997.

[2] Arthur B. Glasser and Gerald E. Subak-Sharpe, *Integrated Circuit Engineering*, Addison-Wesley, Reading, MA, 1979.

[3] Denis Galipeau, *CMOS Latch-up in the D649 Motorola Controller IC*, Cherry Semiconductor, May 9, 1996.

[4] Ronald Troutman, *Latchup in CMOS Technology: The Problem and Its Cure*, Kluwer Academic Publishers, 1986.

[5] Neil Weste and Kamran Eshraghian, *Principles of CMOS VLSI Design*, Addison Wesley, 1985.

[6] S. P. Weeks, *Solid State Tech.* 24, November, 1981, pp. 111-117.

[7] O. D. Trapp, Larry J. Loop, and Richard A. Blanchard, *Semiconductor Technology Handbook*, 6th edition, Technology Associates, Portola Valley, CA, pp. 7-15.

[8] William F. Davis, *Analog I.C. Layout Design Considerations*, Motorola Semiconductor Sector, Mesa, AR, 1981, p. 86.

[9] Giuseppe Massobrio and Paolo Antognetti, *Semiconductor Device Modeling with SPICE*, McGraw-Hill, 1988, Equation 2-143, p. 105.

[10] Cherry *GENESIS Semicustom Design Manual*, pp. 3-25, Cherry Semiconductor Corporation.

chapter 9

Design Practices

Component matching and the protection from electrostatic discharge are important design practices. Accurate component matching reduces costs and improves circuit function. Protection from electrostatic discharge is a necessary precaution for reliability. Often chips are required to pass the human body model electrostatic discharge test discussed in this chapter.

9.1 Matching

While the absolute values of device parameters are difficult to maintain, two devices can be accurately matched in a given circuit. This permits circuit design techniques to be used that result in accurate functions. In this section, chip layout for accurate matching of components is discussed.

Precise matching of components extends performance limits of circuits such as accurate voltage regulators or operational amplifiers with low input offset voltage. Laser trimming or zener zaping can extend performance limits but at the expense of test time and increased die area. Careful attention to matching can improve circuit performance, reduce costs and increase design success.

9.1.1 Component Size

Edge irregularities become a significant fraction of device geometries for small-sized devices. Increasing size reduces the percent variability between matched components.

The same unavoidable edge irregularities exist in large geometry devices as in small. However, as shown in Figure 9.1 for large geometries, the percent variation due to edge irregularities is smaller. Therefore, two large devices will match better than two small devices. If devices are very large, the effects of lateral gradients cause the benefits of large devices to diminish.

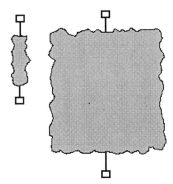

Figure 9.1   Edge irregularities represent a larger fraction of device dimensions in small-sized devices.

9.1.2   Orientation

Matching improves when components are located close together and have the same orientation. This minimizes mismatch due to lateral process variations.

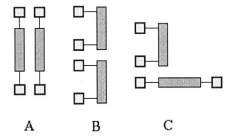

A   B   C

Figure 9.2   The resistors shown in A are oriented for the best match. C represents the worst orientation.

The best match components should be identical, the same size and shape, close together and oriented in the same direction. Lateral process variations such as diffusion gradients, temperature gradients, and mask misalignments will cause component mismatch.

Placing components as close as possible and orienting them in the same direction is the best defense against lateral variations. The layout A in Figure 9.2 will provide the best match. Variations in the Y direction have no effect and the effects of variations in the X direction are minimized by the close proximity of the two devices. In B, variations in X have no effect on matching, but variations in Y do. Since the separation is greater than in layout A, the mismatch will be greater. Layout C is the worst case. Variations in both X and Y effect the match.

9.1.3   Temperature

The presence of a power dissipating component on the chip affects matching. A large resistor or transistor dissipating power causes temperature gradients on the chip. Junction temperatures in power dissipating components can be several degrees above the temperature of the case . Bipolar transistor saturation current I_s is strongly temperature dependent. The simulation described in Section 8.3.2 shows I_s changing by an order of magnitude for a 20-degree change in temperature. Power dissipation and the resultant temperature gradients can be time varying making the behavior of the matched components difficult to understand.

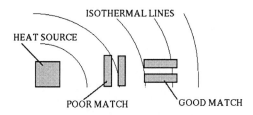

Figure 9.3   Locating matched components equal distance from power dissipating components improves matching.

Locating matched components equidistance from a component dissipating large amounts of power, as shown in Figure 9.3, improves matching.

9.1.4   Stress

Mismatch is greater in packaged dies than in unpackaged wafers. This is due to crystalline stress introduced by the packaging process. For {111} plane wafers, locating matched pairs about the axis of symmetry in the <211> direction near the die center improves matching.

Silicon is a piezoelectric crystal. Stress affects electrical parameters. Chips are packaged at high temperature using materials having thermal coefficients different than silicon. When the package cools to room temperature, stress gradients upset matching.

Wafers used in the bipolar process are cut in the {111} plane. The biCMOS process uses wafers cut in the {100} plane. Figure 9.4 shows a typical packaged bipolar die. Stress is symmetrically distributed about

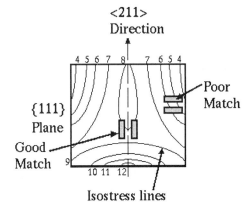

Figure 9.4   A packaged die is shown. Stress is symmetrically distributed about an axis of symmetry in the <211> direction.

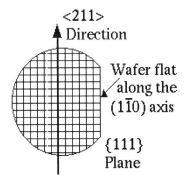

Figure 9.5   In the {111} plane the <211> direction is parallel to the wafer flat edge.

an axis of symmetry, passing through the die center in the <211> direction. Stress gradients are greater near the edges and less near the center. The absolute stress may be greater near the center, but gradients cause mismatch, therefore components placed near the center will match better than components placed near the die edge. For best match, components should be placed to maintain the symmetry, near the die center and equidistant from the axis of symmetry.

In one case, comparisons between measurements taken at wafer probe with measurements taken on packaged circuits show 50% of opamp input offset drift nonlinearity due to packaging.

9.1.5 Contact Placement for Matching

Layout geometries that result in device mismatch when ohmic contacts shift relative to the device, such as a resistor horseshoe layout, should be avoided.

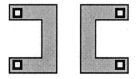

Figure 9.6 This horseshoe layout for matched resistors is to be avoided.

When placing contacts, care has to be taken to minimize the effect of mask layer shifts on matching. Consider the matched pair of resistors shown in Figure 9.6. If contacts shift horizontally, one resistor increases while the other decreases. Vertical shifts have the same effect on both resistors. This horseshoe layout for matched resistors is to be avoided due to the mismatch produced when the contacts shift relative to the resistor.

9.1.6 Buried Layer Shift

Alignment marks consisting of depressions bounded by steps $500A°$ to $100A°$ high are placed on the substrate before the epitaxial layer is grown. The buried layer is aligned to these marks. The buried layer is diffused into the substrate prior to the growing of the epitaxial layer. As the epi is grown, alignment marks placed on the substrate replicate themselves in the epi. Due to anisotropic growth of the epi, alignment marks shift. Since the buried layer is aligned to the mark on the substrate and the other layers are aligned to the shifted mark on the surface of the epi, there is an apparent shift of the buried layer as shown in Figure 9.7. This can influence transistor properties causing mismatch.

S. P. Weeks [1] studied pattern shift during CVD Epitaxy on (111) and (100) silicon. He found typical relative shifts for (111) silicon of 0.6 and a very small shift typically for (100) silicon. Relative shift is defined as the shift divided by the epi thickness. A relative shift of 0.6 represents a shift of 4.8 microns for an epi thickness of 8 microns.

Transistor performance is affected when the edge of the shifted buried layer intersects the emitter or the deep N diffusion.[3] This can affect matching. The effect of buried layer shift is similar to the effect of relative shift of masks. Careful attention to layout with the symmetric

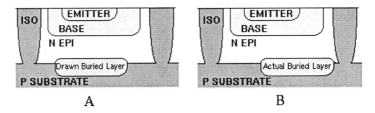

Figure 9.7   A: Drawn - B: Actual. The lateral shift to the right of the actual buried layer relative to the base and emitter is due to the shift to the left of the base and emitter alignment mark. Here the shift causes the edge of the buried layer to intersect the emitter. This can seriously change the effective saturation current [3].

layout of transistors minimizes buried layer shift effects.

Pattern shift is reduced by modifying processing [2] using:

- Higher temperatures to obtain more isotropic growth

- Lower deposition rates

- Lower pressure [this approach can cause faceting (development of undesired crystal planes) or distortion on (100) silicon]

- SiH_4 instead of $SiCl_4$

- (100) silicon, rather than (111) silicon. (Pattern shift of (111) silicon is reduced by cutting the wafer a few degrees off the exact (111) plane)

9.1.7   Resistor Placement

Base resistors in epi tubs are influenced by adjacent diffusions. Two resistors, one adjacent to the isolation well and the other surrounded by other base resistors, will have slightly different resistance values and therefore will not match well.

Diffusions adjacent to a resistor can cause mismatch. The resistors shown in Figure 9.8 are formed using base diffusion in n-epi. Resistors R2 and R3 both have other base resistors beside them. This provides the symmetrical environment required for good matching. Resistors R1 and R4 have isolation well diffusions next to them. They will not match well with R2 and R3. Matching can be improved by adding dummy resistors to assure matching resistors are in identical environments. The dummy resistors may be used for other functions. In Figure 9.8, R1 and R4 are dummies, added to assure R2 and R3 have identical environments.

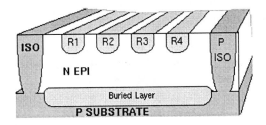

Figure 9.8 The four resistors shown are identical except for their surroundings. Resistors R2 and R3 have symmetrical environments and match. Resistors R1 and R2 do not match well.

9.1.8 Ion Implant Resistor Conductivity Modulation

Lightly doped ion implant resistors are affected by metal passing over them. The potential difference between the resistor and the metal influences the carrier concentration in the resistor and therefore the resistance.

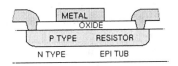

Figure 9.9 Metal passing over p-type ion implant resistors in n-epi is shown.

Figure 9.10 Resistor match is upset by metal passing over one resistor.

Metal passing over ion implant resistors form the metal oxide silicon (MOS) structure. A positive voltage on the metal relative to the resistor repels holes from the surface of the p-type ion implant resistor. This increases the resistivity. The structure is like a MOS transistor with the metal acting as the gate. The metal voltage changes (modulates) the resistor current. Figure 9.9 shows the ion implant resistor with metal passing over it. Figure 9.10 shows matched resistors where the match is upset by metal modulation one resistor.

9.1.9   Tub Bias Affects Resistor Match

The voltage of a resistor relative to its epi tub influences the resistance. Two resistors in the same tub will mismatch if they are at different voltages as, for example, in a voltage divider.

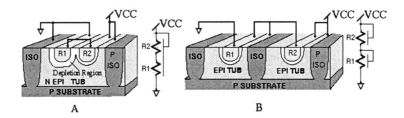

Figure 9.11   A. Different resistor-tub bias produces different depletion regions, changing resistance and upsetting resistance matching. B. Separate tubs biased at the resistor high voltage reduce resistor-tub bias differences improving matching.

The pn junction formed by p-type resistors in n-type epi tubs are reversed biased to isolate the resistors. It is common practice to bias the tubs at the highest voltage in the circuit (VCC) to assure the resistor tub junction is reversed biased. This can contribute to mismatch when different voltages are applied to the resistors as in the voltage divider shown in Figure 9.11A.

The reverse resistor-tub voltage produces a depletion region devoid of charge carriers in the vicinity of the junction. The depletion region extends both sides of the junction. The depth of penetration varies inversely with doping. Therefore, the depletion region extends further into the lightly doped epi tub than it does into the p-type resistor. The depletion region in the resistor reduces the resistor cross section. This increases resistance. The depth of penetration of the depletion region depends on the reverse voltage across the resistor-tub pn junction. The effect is more pronounced for high resistivity ion implant resistors.

Figure 9.11A shows two matched resistors in a single tub. The tub is biased to the highest voltage in the circuit (VCC). The resistors form a voltage divider where the resistor R2 is at a higher voltage than the resistor R1. This results in different depletion regions and therefore different values of resistance than would be expected.

The matched resistors in Figure 9.11B are placed in different tubs. This permits different tub bias voltages. In Figure 9.11B the tubs are biased at the voltage at the high end of the resistor. This results in similar resistor-tub voltages for both resistors with a resultant improvement in matching.

9.1.10   Contact Resistance Upsets Matching

Contact resistance can upset resistor matching when resistor values differ. Matching of large resistor ratios is improved by composing the larger resistor from segments equal to the smaller resistor.

The value of a resistor is the drawn resistance plus the resistance due to the contacts. $R = R_d + 2R_C$, where each contact introduces a resistance R_C. Contact resistance becomes significant when resistance values are small. Large resistor ratios, where one of the resistor values is small, can be upset by contact resistance.

Consider a resistor ratio R_2/R_1.

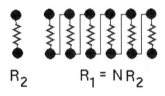

$$R_2 \qquad\qquad R_1 = N R_2$$

Figure 9.12   A large resistance ratio with one resistor composed of multiple repetitions of the smaller resistor achieves a match independent of the contact resistance.

$$\frac{R_2}{R_1} = \frac{R_{d2} + 2R_C}{R_{d1} + 2R_C}$$

If $R_1 \gg 2R_C$

$$\frac{R_2}{R_1} = \frac{R_{d2} + 2R_C}{R_{d1}}$$

The ratio depends on the contact resistance R_C.

A more accurate ratio results when the large resistor is composed of a number of segments, each equal to the value of the smaller resistor as shown in Figure 9.12.

If $R_1 = N(R_{d2} + 2R_C)$

$$\frac{R_2}{R_1} = \frac{R_{d2} + 2R_C}{N(R_{d2} + 2R_C)} = \frac{1}{N}$$

The ratio equals $1/N$, independent of the contact resistance.

9.1.11   The Cross Coupled Quad Improves Matching

A cross coupled quad layout reduces mismatch in the presence of lateral variations. Breaking a component into four parts and laying them out

so opposites are linked reduces mismatch. Positive variations are can-
celed by negative variations in the presence of linear gradients in process
parameters.

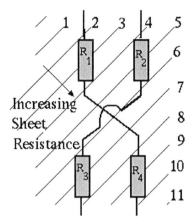

Figure 9.13   Matched quad coupled resistors are shown in the presence of
a linear variation in sheet resistance. $R_1 + R_4$ matches $R_2 + R_3$.

Cross coupled quad resistors are shown in Figure 9.13. A linear gradient in the sheet resistance causes values to vary. R_1 plus R_4 is matched
to R_2 plus R_3. The isoclines represent constant values of sheet resistance in arbitrary units. Resistance values are proportional to the sheet
resistance. The sheet resistances at R_1 and R_4 are 3 and 9 totaling 12.
This matches the total of the sheet resistances at R_2 and R_3 (5 and 7).

Lateral variations of other parameters such as junction depth, oxide thickness, temperature, and stress can be compensated using cross
coupled quads. The technique is not limited to resistors. Matching of
transistors, diodes, and capacitors also benefits from the cross coupled
quad structure.

The nonlinear component of parameter gradients is not compensated
for by the cross coupled quad. For linear gradients, where the spacing
between isoclines is constant, matching is good. When isocline spacing varies, representing nonlinear gradients in the process parameters,
matching is improved but not as much.

9.1.12   Matching Calculations

The performance of precision circuits can be predicted if the accuracy
with which matched components track is known. In this section simple hand calculations of approximate values for amplifier input offset

voltage and gain, and precision quantities that depend on matching are illustrated.

Resistor Matching Calculation

Resistance depends on fabrication tolerances and temperature. The temperature dependence of resistance is to a first order described by α

$$R = R_0(1 + \alpha[T - T_0]) \tag{9.1}$$

where T_0 is room temperature and T is the operating temperature of the resistor. α is called the temperature coefficient of resistance (TCR). The total variation in resistance due to temperature and fabrication variations is

$$\Delta R = \frac{\partial R}{\partial T}\Delta T + \frac{\partial R}{\partial R_{s0}}\Delta R_{s0} + \frac{\partial R}{\partial n}\Delta n \tag{9.2}$$

where the first term is due to temperature variations, the second is due to variations in the sheet resistance and the third is due to variations in n, the resistor geometry. n, the geometric aspect ratio, is the resistor length divided by its width $n = L/W$. For the common case where $L \gg W$, variations in n are dominated by variations in W. The resistor value is the geometric aspect ratio multiplied by the sheet resistance $R = nR_{s0}$:

$$\frac{\partial R}{\partial n} = R_{s0}\left(\frac{\partial n}{\partial W}\Delta W + \frac{\partial n}{\partial L}\Delta L\right)$$

$$= R_{s0}\left(\frac{\Delta L}{W} - \frac{L}{W}\frac{\Delta W}{W}\right)$$

$$\approx -R_{s0}\frac{L}{W}\Delta W = -R\frac{\Delta W}{W} \tag{9.3}$$

Substituting this into Equation 9.2, using Equation 9.1 yields

$$\frac{\Delta R}{R} = \frac{\alpha T}{1 + \alpha(T - T_0)}\frac{\Delta T}{T} + \frac{\Delta R_{s0}}{R_{s0}} - \frac{\Delta W}{W} \tag{9.4}$$

Variations in temperature, variations in process parameters, as reflected in sheet resistance, and variations in resistor geometry affect resistor values and therefore matching. If two nominally identical resistors are laid out to minimize differences in temperature and process variations, random variations in resistor width remain causing resistor mismatch.

While the absolute values of the resistors vary widely, matched resistors will track to within a fraction of a percent. Precise circuit functions

are realized using ratios of matched resistors. For two nominally identical resistors R_1 and R_2, their ratio is

$$\frac{R_1}{R_2} = \frac{R_1 + \Delta R_1}{R_2 + \Delta R_2} = \frac{1 + \frac{\Delta R1}{R_1}}{1 + \frac{\Delta R_2}{R_2}}$$

$$= \frac{1 - \frac{\Delta W_1}{W_1}}{1 - \frac{\Delta W_2}{W_2}} \approx 1 - \frac{\Delta W_1 - \Delta W_2}{W} = 1 - \frac{2\Delta W}{W} \tag{9.5}$$

There is a tradeoff between resistor size and matching, the larger W the
better the match. But W cannot be increased indefinitely. At some
point, process parameter gradients become important and the variation
of sheet resistance with distance will become a factor.

Example

If the two resistors shown in Figure 9.14 track so that their ratio is
off by 0.1%, by what percent will the output voltage be off?

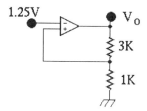

Figure 9.14 A bandgap regulator output of 1.25 volts is multiplied to
achieve a 5 volt regulator output, using an opamp and a precision voltage
divider. Resistor mismatch affects output voltage V_o.

Answer

If the opamp in Figure 9.14 is ideal:

$$V_o = V_{bg}\left(1 + \frac{R_1}{R_2}\right)$$

In this example, $V_{bg} = 1.25$ V, and $R_1/R_2 = 3[1 \pm .001]$, (for a 0.1%
error in the resistor ratio). Therefore,

$$V_o = 1.25[1 + 3(1 + .001)] = 5[1 \pm .00075]$$

This represents a 0.075% variation in the output voltage.

Transistor Mismatch Calculations

Consider the bipolar comparator shown in Figure 9.15. Transistor mismatch results in transistors with different saturation currents. If the two collector currents I_{N1} and I_{N2} are to be the same, an input offset voltage must be applied to the bases of the transistors N1 and N2. In

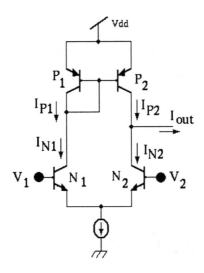

Figure 9.15 Transistor mismatch influences comparator input offset voltage.

Figure 9.15,

$$I_{N1} = I_{c1} = I_{s1} exp \left(\frac{V_{be1}}{V_T} \right)$$

$$I_{N2} = I_{c2} = I_{s2} exp \left(\frac{V_{be2}}{V_T} \right)$$

$$\frac{I_{N1}}{I_{N2}} = \frac{I_{s1}}{I_{s2}} exp \left(\frac{\Delta V_{be}}{V_T} \right)$$

where $\Delta V_{be} = V_{be1} - V_{be2}$. In general $I_{s1} \neq I_{s2}$ due to transistor mismatch. For equal collector currents, the offset voltage, ΔV_{be} must be applied to the transistor bases.

$$\Delta V_{be} = V_T ln \left(\frac{I_{s2}}{I_{s1}} \right)$$

If the ratio of the saturation currents, I_{s2}/I_{s1}, is off by 20%, the offset voltage, ΔV_{be}, is 4.7 mV.

Now suppose the two transistors N1 and N2 are perfectly matched, but the transistors P1 and P2 are mismatched and produce different currents I_{P1} and I_{P2}. If the output current I_{out} is to be zero, an input offset voltage ΔV_{be} has to be applied to the bases of N1 and N2, so that $I_{N2} = I_{P2}$, then I_{out} will be zero. If the ratio of I_{P1}/I_{P2} is off by 20%, I_{N1}/I_{N2} must also be off by 20% to achieve $I_{N2} = I_{P2}$ and zero I_{out}. This requires an input offset voltage ΔV_{be} of 4.7 mV. In the worst case npn transistor mismatch will add to pnp transistor mismatch. If both transistor pairs mismatch by 20%, an input offset voltage of 9.4 mV could result.

MOS Transistor Mismatch

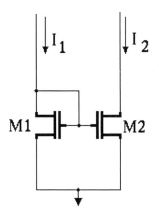

Figure 9.16 If the two transistors are identical, the current I_2 mirrors I_1. However, if the transistors do not match, the currents will be different.

The two source-coupled MOS transistors shown in Figure 9.16 could be part of a comparator or a current mirror. Variations in geometry and threshold voltage result in different drain currents. This is a mismatch. The drain current is

$$I_D = \frac{W}{L} \frac{\mu Cox}{2} [V_{GS} - V_{th}]^2 \tag{9.6}$$

If the geometry and threshold voltages are different:

$$\Delta I_D = \frac{\partial I_D}{\partial \left(\frac{W}{L}\right)} \Delta \left(\frac{W}{L}\right) + \frac{\partial I_D}{\partial V_{th}} \Delta V_{th} \tag{9.7}$$

where from Equation 9.6

$$\frac{\partial I_D}{\partial\left(\frac{W}{L}\right)} = \frac{I_D}{\frac{W}{L}} \tag{9.8}$$

$$\frac{\partial I_d}{\partial V_{th}} = -\sqrt{2I_D\frac{W}{L}\mu Cox} = -g_m \tag{9.9}$$

substituting these results into Equation 9.7

$$\frac{\Delta I_D}{I_D} = \frac{\Delta\frac{W}{L}}{\frac{W}{L}} - 2\frac{\Delta V_{th}}{V_{GS} - V_{th}} \tag{9.10}$$

Since the threshold voltage V_{th} varies with position on the chip, it is important to place matched MOS transistors close together. If the matched transistors form a current mirror with the output current needed at a remote location, it is better to keep the transistors close together and run metal to carry the output current to the remote location.

Example

Calculate the input offset voltage for the circuit in Figure 9.16 if the threshold voltage difference is 0.01 V and the aspect ratio W/L is 10. The aspect ratios for the two transistors differ by 0.1%. The nominal drain current is 100 μA and μCox is 1E-6.

Answer

If the drain current is a function of W/L, V_{th} and the input voltage, V_{GS},

$$\Delta I_D = \frac{\partial I_D}{\partial\frac{W}{L}}\Delta\frac{W}{L} + \frac{\partial I_D}{\partial V_{th}}\Delta V_{th} + \frac{\partial I_D}{\partial V_{GS}}\Delta V_{GS} \tag{9.11}$$

From Equations 9.8 and 9.9 and the definition of g_m:

$$\Delta I_D = \frac{\Delta\frac{W}{L}}{\frac{W}{L}} - g_m\Delta V_{th} + g_m\Delta V_{GS} \tag{9.12}$$

Set ΔI_D equal to zero to find the difference in V_{GS} that will produce the same drain current for different W/L and V_{th}:

$$\Delta V_{GS} = \Delta V_{th} - \frac{I_D}{g_m\frac{W}{L}}\Delta\frac{W}{L} \tag{9.13}$$

- $I_D = $ 1E-4

- $W/L = 10$

- $g_m = \sqrt{2I_D(W/L)\mu Cox} = 4.47\text{E-}5$

- $(\Delta W/L)/(W/L) = 0.001$

- $\Delta V_{TH} = 0.1$ V

Plugging these values into Equation 9.13, assuming the worst case where the effect on V_{GS} add

$$\Delta V_{GS} = \Delta V_{th} + 0.224\Delta\frac{W}{L} = 0.01 + 0.0024 = 0.012 \; V$$

9.2 *Electrostatic Discharge Protection (ESD)*

The gate oxide of MOS transistors breaks down at a few tens of volts. Circuits containing these fragile structures have to be protected from electrostatic discharges (ESD) produced by the human body and mechanical objects charged to thousands of volts and capable of pulses of current in the tens of ampere range.

Electrostatic discharge (ESD) threatens the reliability of integrated circuits. Human Body ESD events are common. The human body capacitance of 150 pF is charged to 4 KV by 0.6 microcoulombs. If a charged person contacts a grounded object, such as an ic pin, discharge currents in the ampere range can result for about 100 ns. Sometimes damage can be too weak to be detected easily, resulting in "walking wounded" or "latency effects."

A device can be exposed to ESD at any point in its lifetime, from the fabrication to end use in a circuit board.

The breakdown field (dielectric strength) of silicon dioxide is about $10^7 \; V/cm$. For a gate oxide thickness of $250°A$. The electric field in this thin oxide equals the breakdown field when the voltage across it is 25 V.

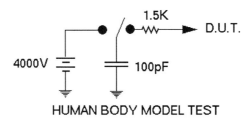

HUMAN BODY MODEL TEST

Figure 9.17   The Human Body Model (HBM).

In order to meet customer specifications, integrated circuits may have to pass the human body model (HBM) test. It is the principle test method. The HBM test is specified in the MIL-STD 883C method 3015.7.[5] An ESD pulse is generated by discharging a 100-pF capacitor through a 1.5 K resistor into the device under test to approximate the discharge from a human body as shown in Figure 9.17.

Machine Model (MM) and Charged Device Model (CDM)

For testing, the machine model (MM) and the charged device model (CDM) of ESD events are sometimes required. The machine model (MM) is a 200pF capacitor discharged through zero ohms. This leads to higher currents limited to 2-3 amperes by parasitic series resistance and inductance. For the charged device model (CDM) the device itself is charged to a few KV. The ESD stress occurs when one pin of the charged device is grounded.

Failure Modes

- In properly designed ESD protection circuits, the weakest links are diffusions.[6] The damage is usually located at the diffusion connected to the pad being stressed (or the drain of the NMOS), and can be localized at the junction edge or extended under the channel and across to the source of the transistor.

- Damage to contacts is rare in advanced process where barrier metals are used between aluminum metalization and the silicon.[6]

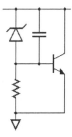

Figure 9.18   ESD current is absorbed by a large transistor triggered by a low voltage zener. Zener capacitor speeds turn on in response to fast ESD transients.

- Oxide damage is becoming more of a concern as gate oxide thickness decreases. According to Duvvury and Amerasekera[6] oxide failures are rare. These failures are usually located in the PMOS of the CMOS input gate and are very often found between the gate and the diffusion connected to the power supply.

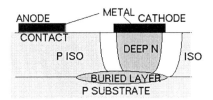

Figure 9.19   The n-type deep buried layer and the p-type isolation (iso) form a pn junction that breaks down at about 12 V and can carry the large ESD currents.

Deep junctions like the buried layer-isolation (BL/ISO) zener shown in Figure 9.19 have smooth rounded edges and can take large currents. Current in surface junctions such as the NSD/ISO zener tends to be nonuniform with high current densities at corners. The attractive breakdown voltage of the NSD/ISO zener (about 6 V) can be used to turn on a large transistor as shown in Figure 9.18.

Protection transistors should be designed with emitter degeneration to provide the debiasing necessary to produce a uniform distribution of current. Long linear layouts avoid the bunching of current that can occur when a finger type layout is used.

9.3   ESD Protection Circuit Analysis

A hand calculation is done to predict the performance of ESD protection circuits. Consider the chip under test in Figure 9.20. A human body model tester, consisting of a 150 pF capacitor and a 1.5 K resistor, is used to test the circuit protecting the MOS gate. The protection circuit includes two zener diodes and a 2 K resistor. Neglecting parasitic capacitances results in a conservative calculation since they shunt current away from the MOS gate. The 150 pF capacitor is charged to the test voltage of 2 KV. The test event begins when the switch closes. At that time, neglecting parasitics, a peak current $I_1 = 2\ KV/1.5\ K\Omega = 1.33\ A$ flows out of the capacitor and into the chip under test. The buried layer/iso zener breaks down and conducts approximately all of this current. In the process, $V_1 = 37.3$ V develops across the BL/ISO zener. The buried layer iso zener is modeled as a 12 V source in series with a 19 Ohm resistor as shown in Figure 9.21.

$$V_1 = 12\,V + 19I_1 = 37.3\,V$$

The bulk of the 2 KV test voltage is absorbed by the 1.5 K resistor (part of the human body model). The remaining 37.3 V is applied to the on-chip 2 K resistor in series with the NSD/ISO zener. The zener

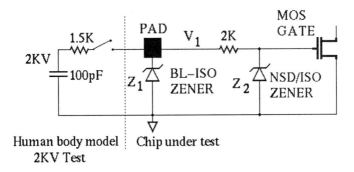

Figure 9.20 ESD Protection circuit being analyzed.

Figure 9.21 The buried layer iso zener acts like a 12 V battery in series with a 19 Ohm resistor.

breaks down at 5 V and is modeled as a 5 V battery in series with a 65 Ohm resistor as shown in Figure 9.22.

$$I_2 = \frac{37.3 - 5}{2\,K} = 15.6\,mA$$

15.6 mA flows through the 2 K resistor and is absorbed by the NSD/ISO zener. In the process, 6 V develops across the zener at the MOS gate. The original 2 KV test voltage has been reduced to 6 V. The MOS gate is saved.

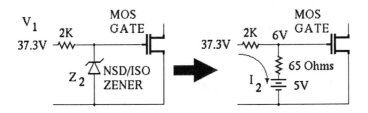

Figure 9.22 The NSD iso zener acts like a 5 V battery in series with a 65 Ohm resistor.

- The MOS gate survived since only 6 V occurred at the gate. The maximum allowed was 12.5 V.

- Z_1 may have not survived if it was not large enough to carry the peak current of 1.3 A in this 2 KV test example.

- If the input is doubled to 4 KV, $I_1 = 2.6\ A$, $V_1 = 61.4\ V$, $I_2 = 27.12\ mA$, and the voltage at the MOS gate is $6.8\ V$.

- Metal may not have survived. Metal widths for this example should be 25 microns per ampere of peak current for ESD events. The ESD current in a 4 KV test is about 3 A. Metal carrying this current should be greater than 75 microns wide.

- The maximum safe gate-to-source voltage for an MOS gate is 5 megavolts per cm of oxide thickness. IF $250A°$ gates are used in low voltage MOS, the gate voltage should not exceed 12.5 V.

- The zeners see the ESD stress as coming from a current source since it comes through a relatively large resistor (1.5 K for the BL/ISO zener and 2 K for the NSD/ISO zener in this example).

- Care has to be taken in scaling zeners. Series resistance of the buried layer/ISO zener scales inversely with zener area. The NSD/ISO zener is a surface structure. Breakdown occurs at a point along the perimeter. Surface zener series resistance does not scale with area.

- Greater protection can be achieved by increasing the 2 K resistor. In normal operation, it is in series with the large impedance of the MOS gate in parallel with an off NSD/ISO zener. Its main effect is to produce an RC time constant that limits high frequency performance. C is the small parasitic gate capacitance (perhaps 0.1 pF).

Example 1

In applications (such as the pad is an I/O pin), where large signal currents must pass through the 2 K resistor producing undesirable voltage drops, the circuit shown in Figure 9.23 permits lower R values. With the assumptions used above, calculate the the peak voltage at the MOS gate if R is 100 Ohms. Will the gate survive if the maximum allowed gate voltage is 12.5 V?

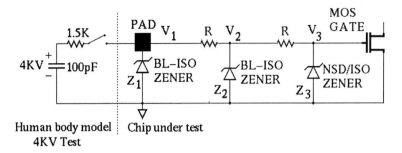

Figure 9.23   Two buried layer iso zeners Z_1 and Z_2 permit the use of lower values of resistance.

Answer

Assuming all the capacitor current flows through the first BL/ISO zener, represented by a 12 V battery and a 19 Ohm series resistance, the voltage across it is

$$V_1 = 12 + \frac{4000V}{1500\ \Omega} 19\ \Omega = 62.7\ V$$

The current through the first 100 Ohm resistor and the second BL/ISO zener is

$$I_2 = \frac{62.7 - 12}{119} = 0.43\ A$$

This assumes the full I_2 flows in the second BL/ISO zener. The voltage across it is

$$V_2 = 12 + 19I_2 = 20\ V$$

The voltage across the NSD/ISO zener, represented by a 5 V battery and a 65 Ohm series resistance is

$$V_3 = V_{MOSGATE} = \frac{20 - 5}{165} 65 + 5 = 11\ V$$

11 V is less than the maximum allowed gate voltage of 12.5 V.

The analysis was conservative since it ignored parasitic capacitance and treated

$$I_3 \ll I_2 \ll I_1$$

The approximate currents through the zeners are

- $I_1 = 2.7\ A$

- $I_2 = 0.43\ A$

- $I_3 = 0.09\ A$

The approximate node voltages are

- $V_1 = 62.7\ V$

- $V_2 = 20\ V$

- $V_3 = 11\ V$

9.4   Chapter Exercises

1. A resistor voltage divider produces an output voltage

$$V_o = \frac{R_1}{R_1 + R_2} V_{in}$$

   If $V_{in} = 5\ V$, $R_1 = 3\ K\Omega$ and $R_2 = 2\ K\Omega$, what is the output voltage? If the resistor values are 20% larger, what is the output voltage? If R_1 is 20.1% larger and R_2 is 19.9% larger, what is the output voltage? What is the percent error in the output voltage?

2. The transistor pair shown in Figure 9.24A have differences resulting in a 20% difference in their saturation currents I_s. What difference in base voltages will produce the same collector currents?

3. The opamp circuit shown in Figure 9.24B is used to multiply up a 1.25 V reference. R1 is 1.4 KΩ. R2 is 1 KΩ. What is the output voltage? If the resistor values are 20% low, by what percentage does the output change? If the resistor ratio is off by 0.1%, what is the percent error in the output voltage?

4. What is the worst case input offset voltage required to keep the output current I_{out} zero for the amplifier shown in Figure 9.15 if the saturation currents I_s for the npn transistors N1 and N2 are mismatched by 10% and the saturation currents for the pnp transistors P1 and P2 are mismatched by 20%?

5. The D/A converter shown in Figure 9.25 consists of binary weighted transistors. The transistor length L is the same for all transistors. The widths are binary weighted $W1 = 2W0$, $W2 = 4W0$, $Wn = 2^n W0$, etc. The input voltages $V0$, $V1$ etc. are 0 or 5 V. The output current is proportional to the input word represented by the voltages $V0$ through Vn. The nominal threshold voltage is 1 V. The mismatch can be attributed to a 0.01 V uncertainty in the threshold.

   a. How many bits can be accurately converted?

   b. What is the percent error in the output current?

6. If the ESD protection circuit shown in Figure 9.23 is composed of BL/ISO zeners with series resistance of 30 Ohms and a turn on voltage V_z of 10 V and NSD/ISO zeners with series resistance of 130 Ohms and V_z of 5 V, what will the gate voltage be when the pad is subjected to a 4 KV human body model stress test?

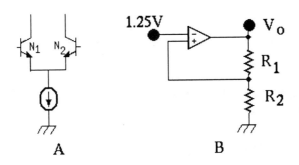

Figure 9.24   For these problems matched transistors and resistors differ by 20%.

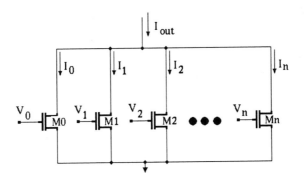

Figure 9.25   Digital to analog converter, DAC, formed using transistors with binary weighted widths. The output current is proportional to the binary input word.

References

[1] S. P. Weeks, *Solid State Tech.* 24, November, 1981, pp. 111-117.
[2] O. D. Trapp, Larry J. Loop, and Richard A. Blanchard, *Semicon-*

ductor Technology Handbook 6^{t}h edition, Technology Associates, Portola Valley, CA, pp. 7-15.

[3] William F. Davis, *Analog I.C. Layout Design Considerations*, Motorola Semiconductor Sector, Mesa, AR, 1981, p. 86.

[4] Charvaka Duvvury and Ajith Amerasekera, *State-of-the-art issues for technology and circuit design of ESD protection in CMOS ICs*, Semiconductor Science and Technology, June 96, pp. 833-850.

[5] EOS/ESD *Standards No. 5.0 Human Body Model(HBM)* Draft, November 1988, Electrostatic Discharge Testing, EOS/ESD Association, Inc., 201 Mill Street, Rome, New York 13440.

[6] Duvvury C., Amerasekera A., *1993 ESD: a pervasive reliability concern for IC technologies,* Proc. IEEE **81**, pp. 690-702

[7] Amerasekera, A., Hannemann, M., and Schofield, P., *1992 ESD failure modes: characteristics, mechanisms and process influences,* IEEE Trans. Electron Devices **39**, pp. 430-436

Index

acceptor 3
amplifier
   common-base 89, 96, 99
   common-collector 89, 99
   common-drain 89, 117
   common-emitter 89, 90, 99
   common-gate 89
   common-source 89, 113
   common-mode 107
   common-mode gain 109, 111, 121
   common-mode half-circuit 111, 121
   common-mode input range 112
   differential 104, 107
   differential gain 108, 111, 121
   differential half-circuit 109
   differential input 105
   input voltage range 175
   MOS cascode 116
avalanche breakdown 12
avalanche multiplication 26
bandgap 85, 160, 204
bandgap voltage 83
barrier lowering 46
beta 21, 22, 25, 49, 167
body effect 54, 73
body transconductance 54
breakdown 13, 25
breakdown voltage 13
BSIM 45
buffered divider 80

   temperature compensated 80
built-in potential 11
buried layer 20
buried layer shift 197
BV_{CBO} (*see* breakdown)
BV_{CEO} (*see* breakdown)
cascode 89, 99, 103
cascoding 96
CE-CB amplifier 103
channel length modulation 46
channel stops 189
charge control 42
charge neutral region 10
charge sharing 47
charged device model 209
CMRR (*see* common-mode rejection ratio)
CMOS 30
   inverter 89, 114
collector current temperature dependence 171
common-mode rejection ratio 111, 121
comparator 173-174
conduction band 1
conductivity 1
conductivity modulation 199
contact potential 30
contact resistance 201
cross coupled quad 201
curent mirror 58, 162, 163
   cascode 68
   output resistance 71

Widlar 67
Wilson 69
current gain 61
current gain beta 25
current source 169
   V_be/R 71
   $\Delta V_be/R$ 71
   Widlar 65
Darlington 89, 101, 173
depletion region 10, 200
DIBL (*see* drain induced barrier
    lowering)
diffusion capacitance 16, 44
diffusion current 8, 17
diode equation 59
donor 3
drain induced barrier lowering
   46
drift velocity 4, 47
Early effect 39, 43, 49
Early voltage 50, 52, 66
Ebers-Moll model 22, 25
electrostatic discharge 208
   protection circuits 210
emitter coupled pairs 89, 104
emitter degeneration 68
emitter resistors 164
emitter follower 89, 98
epiFETs 35
ESD (*see* electrostatic discharge)
excess holes 16
Fermi level 30
gm 52, 53
Gummel plot 44, 49
Gummel-Poon model 40
headroom 173
high level injection 39, 40, 43,
   44, 49
hole 1, 2, 6, 16
hot carrier effects 48
human body model 208
hysteresis 174
ideality factor 49
incremental capacitance 14

injection 16
intrinsic silicon 2
ion implant resistor 190
ionized atom 2
ir drop 159, 162, 163
junction capacitance 14, 44
junction field effect transistor 35
Kelvin line 160-162
latchup 179-183
lateral pnp 166, 167
law of mass action 3, 6
law of the junction 15, 20
machine model 209
majority carrier 3
matching 193
minority carrier 3
mobility 4, 169
mobility variation 48
MOS cascode amplifier 116
MOS cascode structures 73
MOS model 31
n-type 3
offset voltage 112, 122
operational amplifiers 105
OSFET 187, 188
output impedance 52, 67, 68
pn junctions 9
protection circuit 210
PTAT 84
punch through 12, 13
rectifier equation 19
resistor
   temperature dependence 203
   ion implant 190, 199
saturation 45
saturation current 19, 49
   temperature dependence 171
Schottky diode 184
SCR 183
sheet resistance 7
silicon controlled rectifier 183
small signal models 51, 53
source flapping 190
source-coupled pair 89, 118

space-charge region 10
SPICE 49
stored charge 177
stress 195
surface states 48
tail current 173
thermal voltage 165
transconductance 52, 53
    body 54
transistor
    bipolar 19
    composite 99, 100
    jfet 35
    MOS 26, 44
    MOS model 31
    MOS linear region 45
    MOS mismatch 206
    MOS saturation 33
    MOS small signal model 53
    MOS threshold voltage 54
    parasitic 185, 187
    parasitic pnp 166, 169, 185
    pnp beta 167
    pnp saturation 166
    SPICE model parameters 50
    vertical npn 19
tub bias 200
Vbe multiplier 80
velocity saturation 47
voltage reference
    bandgap 85
    ideal 79
    zener 82
zener diode 34, 82, 209, 210
    buried layer iso zener 34